AQA ... tics

Revision Guide

New GCSE

Tony Fisher
June Haighton
Andrew Manning
Anthony Staneff
Margaret Thornton
Series Editor
Paul Metcalf

W... ...nes

Published in 2010 by:
Nelson Thornes Ltd
Delta Place
27 Bath Road
CHELTENHAM
GL53 7TH
United Kingdom

10 11 12 13 14 / 10 9 8 7 6 5 4 3 2 1

A catalogue record for this book is available from the British Library

ISBN 978 1 4085 0620 2

Cover photographs by iStockphoto (top), PureStock/Photolibrary (middle), Photolibrary (bottom)
Illustrations by Tech-Set Limited
Page make-up by Tech-Set Limited, Gateshead

Printed and bound in Spain by GraphyCems

Photograph acknowledgements
iStockphoto: p32, p41, p46.

Contents

How to use this Revision Guide

This book has been written by teachers and examiners to prepare you for your AQA GCSE Mathematics exams. It covers all the main points you need to know, and it includes references to the Student Books if you want greater detail. If you are taking the modular course you will be assessed in each of three units. The signpost icon Unit 1 is used throughout this guide to show you what unit the content will be assessed in. If you are taking the linear course, you will be assessed on any of the content in each of two exams, one of which will be non-calculator. The following icon is used throughout this guide to show that you should answer a question without using a calculator:

There are some free, downloadable resources available to accompany this guide. See the back cover for details.

In the exams, you will be tested on the following Assessment Objectives (AOs):

AO1 recall and use your knowledge of the prescribed content

AO2 select and apply mathematical methods in a range of contexts

AO3 interpret and analyse problems and generate strategies to solve them.

You will also be assessed on your Quality of Written Communication (QWC).

This book is split into five sections:

- Number – all the basic number work you need to know. Remember that for some of the work you won't be able to use your calculator.
- Statistics – all this material is assessed in Unit 1 if you are taking the modular course.
- Algebra – this is covered in Unit 2 and Unit 3 in the modular course. As with the number work, some of this work must be done without using a calculator.
- Geometry and measure – all this material is assessed in Unit 3 if you are taking the modular course.
- Essential skills – this section looks at the skills you will need to cover QWC, AO2 and AO3 (see the table above).

Each chapter has the following features:

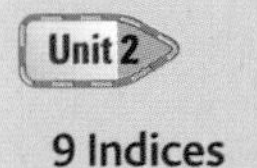

This feature shows you where in the Student Books the material is covered if you need to look at it in more detail.

Revise... Key points

These are the basic points that you need to understand for the topic.

Key terms

Make sure that you know the meaning of the words in this list by writing down your own definitions and checking them with the glossaries in the Student Books. You are expected to use specialist vocabulary as part of your Quality of Written Communication.

Example

This is an example and worked solution to help show you how to use what you have just revised.

Practise...

As in the Student Books, there are questions that allow you to practise what you have just revised.

The bars that run alongside questions in the exercises show you what grade the question is aimed at. This will give you an idea of what grade you are working at. Do not forget: even if you are aiming at a Grade A, you will still need to do well on the Grades D–B questions.

These questions are Functional Maths type questions, which show how maths can be used in real life.

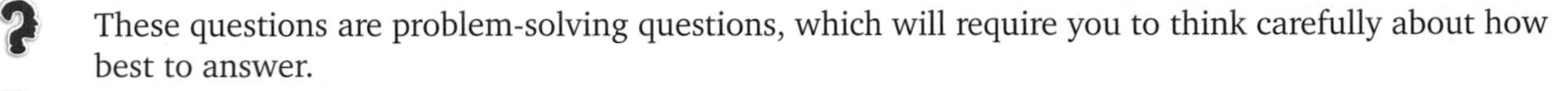

These questions are problem-solving questions, which will require you to think carefully about how best to answer.

These questions should be attempted with a calculator.

These questions should be attempted without using a calculator.

AQA Examination-style questions

At the end of each section there are some questions in the style of those that you will meet in your exams.

Hint

These are tips for you to remember while learning the maths or answering questions.

AQA Examiner's tip

These are tips from the people who will mark your exams, giving you advice on things to remember and watch out for.

Bump up your grade

These are tips from the people who will mark your exams, giving you help on how to boost your grade. These are especially aimed at getting a Grade C.

Aim higher

These are tips from the people who will mark your exams, giving you help on how to boost your grade, especially aimed at getting a Grade A/A*.

Revision tips

WHAT

- Ask your teacher which specification you are following and download a copy from the web. Use this Revision Guide and your exam specification to see what topics you need to cover.
- Go through the topics, past papers and tests and make a list of the areas that you find difficult. Concentrate your revision on these areas.
- Check out which formulae will be given on the exam paper. Your teacher can tell you this. Practise using these and make sure that you learn any formulae that aren't given.

HOW

- Create an effective revision timetable. A blank timetable spreadsheet is available in the downloadable resources.
  - Find out the date of your exam.
  - Divide your subjects up into topics.
  - Mix and match harder and easier topics to break it up a bit.
  - Schedule 10 minutes of top-up time at the start of each session to look back over the work you covered last time.
  - Tick off topics as you complete them. Some checklists are provided in the downloadable resources.
  - Timetable in some time off and rewards.
  - Allow time in the last week to do examination-style questions and go over everything one last time.
  - Stick to your schedule, but if you feel comfortable with some topics and are struggling with others, shift it around to allow extra time on the harder topics.
- Use cards to summarise the key points listed in this book. Condense them to one side of paper and take it everywhere with you, reading it at every opportunity.
- Create mind maps or spider diagrams for different topics using plenty of colour, and stick them on your wall. A blank spider diagram is provided in the downloadable resources.
- Don't try to memorise maths; try to understand the processes.
- Use the end-of-chapter and examination-style questions to practise, practise, practise.

Exam tips

- Take the correct equipment: two pens (only write in blue or black), two sharp pencils, a pencil sharpener, a ruler, an eraser, your calculator with spare batteries, protractor, compasses and a watch to time your answers.
- Things you need to know about **your** calculator before you sit your exam:
  - check that your calculator follows BIDMAS rules (try typing $20 - 9 \times 2 =$; if your calculator gives the answer 2 it is following the rules, if it gives the answer 22 it is not and it is worth getting a newer calculator)
  - how to use brackets
  - how to use the memory
  - how to use the Ans button to insert the previous answer (not all calculators have this function)
  - how to find a square root ($\sqrt{\ }$)
  - how to find a power of a number (3^2, 5^4, ...)
  - how to do fractions
  - how to select degree mode
  - how to enter and read figures in standard form.
- Lower-graded, easier questions are at the beginning of the paper, so start with those to ease your way in.
- Pace yourself so that you do not run out of time, and try to allow 10 minutes at the end for checking your answers.
- Try to work neatly and set out your answer clearly in the given space.
- Look at the marks available for a question. A question worth 1 mark does not need lots of explanation or working, but a 4-mark question is likely to need more than one step to get the answer.
- Show all of your working for questions worth more than 1 mark. If you simply write down the answer, you are taking a chance that it is completely correct. The examiner will not be able to award you part marks when your method is not shown.
- Use appropriate calculator methods on the calculator paper. For example, build up and long multiplication are methods that are best suited to the non-calculator paper.
- Read the question carefully and interpret the 'exam speak':

Exam speak	What it means
Write down	Working is not needed.
Show; You **must** show your working; Explain/Justify/Support your answer; Give a reason for your answer	Show your method; you will not get any marks otherwise. You can use words, numbers or algebra.
Estimate	Round the numbers, then do the calculation. Do not find the exact answer.
Work out/Calculate/Find	Do a calculation; do not measure.
Measure	Use a ruler or protractor; do not calculate.
Not drawn accurately	The diagram is not accurate; do not measure it.
Use the graph to estimate/solve	Your answer must come from the graph.
Give your answer to a suitable degree of accuracy	Use the same or less accuracy than the numbers given in the question.
State the units of your answer	There will be a special mark for this.

- Check that your answer is sensible. For example, if it is a probability, is it between 0 and 1?
- If the question asks you to give your answer to a degree of accuracy, then make sure you do.
- Do not round until the end of the question (unless it is an estimating question) or you may lose accuracy.
- Study the examination-style answers to see what the examiner is looking for.

GOOD LUCK and DON'T PANIC!

Number

1 Numbers and surds

Unit 2

1 Prime factors

4 Surds

Key terms

Write down definitions for the following words. Check your answers in the glossary of your Student Book.

common factor

factor

highest common factor (HCF)

index

irrational number

least common multiple (LCM)

multiple

prime number

product

rational number

surd

Revise... Key points

Note that you are expected to be able to do all of the work in this chapter without a calculator.

Factors and multiples Unit 2

To find all the **factors** of a number, look for factor pairs. For example, the factors of 20 are 1, 2, 4, 5, 10, 20

The **product** of each pair is 20.

To find the highest common factor of two numbers, list all their factors. The **highest common factor (HCF)** is the biggest number that appears in both lists.

To find the **least common multiple (LCM)** of two (or more) numbers, list their **multiples**. The least common multiple is the smallest number that appears in both lists.

Bump up your grade

For a Grade C, you should be able to find highest common factors and least common multiples.

Prime factors Unit 2

You can find the prime factor decomposition of difficult numbers using the tree method or the ladder method. Remember that 1 is **not** a prime number.

Tree method

One of the ways of splitting up 540 is shown below.

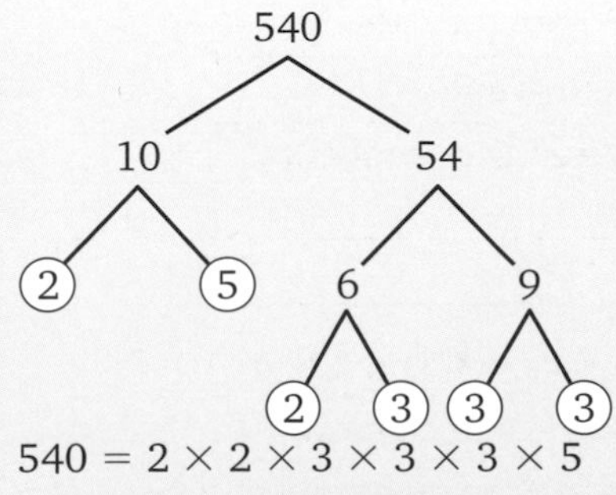

Keep splitting up the numbers until you reach prime numbers.

$540 = 2 \times 2 \times 3 \times 3 \times 3 \times 5$

In **index** notation $540 = 2^2 \times 3^3 \times 5$

Alternatively, you could start the tree with the factors 2 and 60. (Remember that 2 is a factor of all even numbers.)

Ladder method

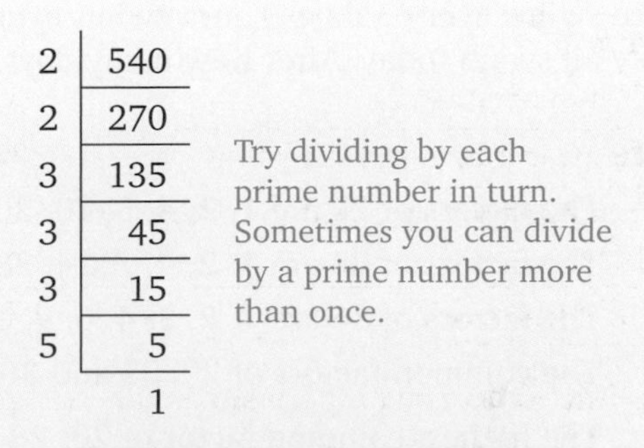

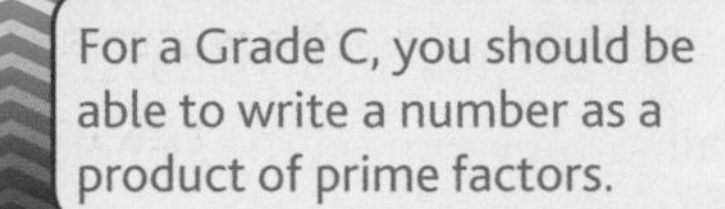

Bump up your grade

For a Grade C, you should be able to write a number as a product of prime factors.

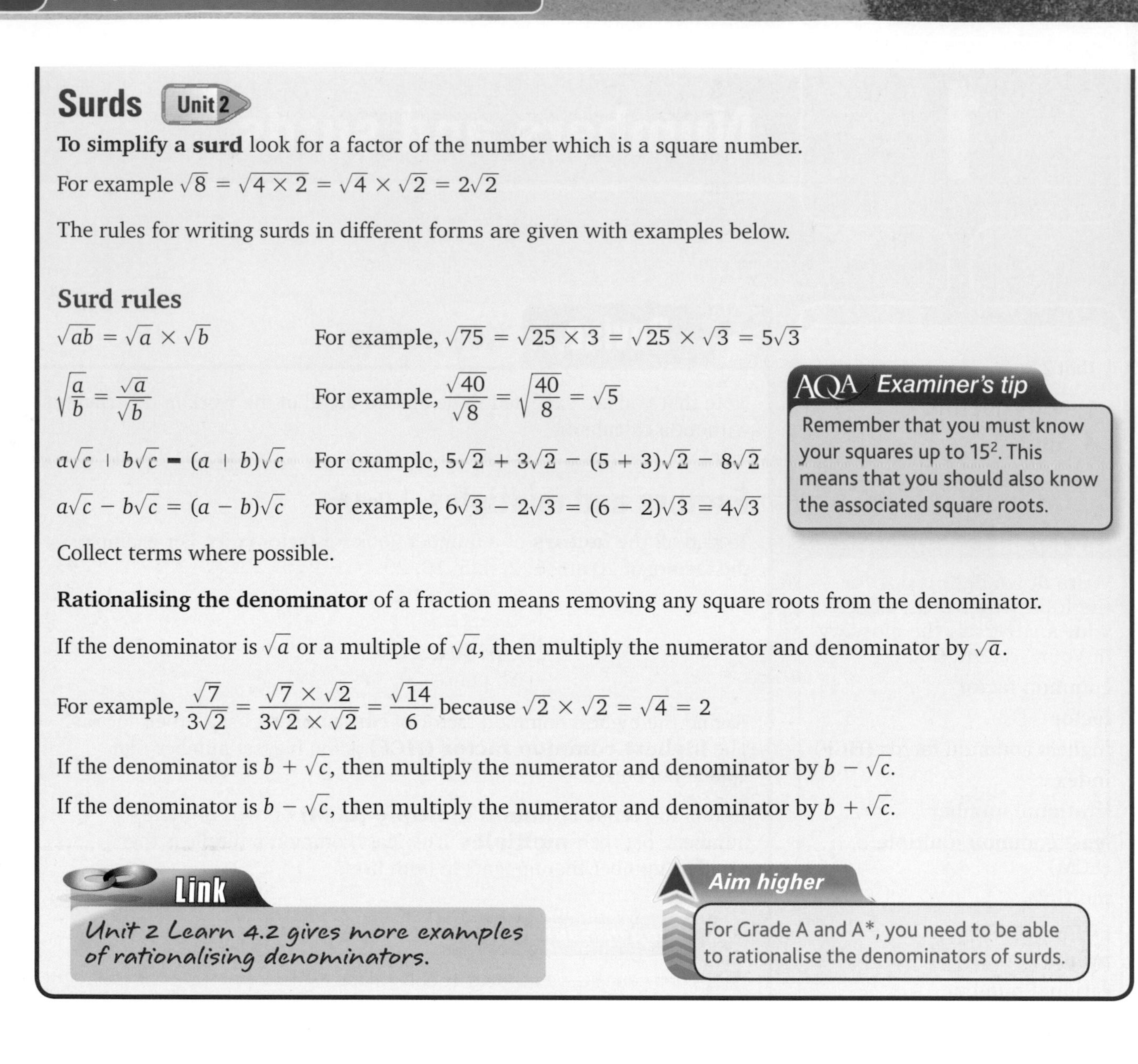

Surds Unit 2

To simplify a surd look for a factor of the number which is a square number.

For example $\sqrt{8} = \sqrt{4 \times 2} = \sqrt{4} \times \sqrt{2} = 2\sqrt{2}$

The rules for writing surds in different forms are given with examples below.

Surd rules

$\sqrt{ab} = \sqrt{a} \times \sqrt{b}$ For example, $\sqrt{75} = \sqrt{25 \times 3} = \sqrt{25} \times \sqrt{3} = 5\sqrt{3}$

$\sqrt{\frac{a}{b}} = \frac{\sqrt{a}}{\sqrt{b}}$ For example, $\frac{\sqrt{40}}{\sqrt{8}} = \sqrt{\frac{40}{8}} = \sqrt{5}$

$a\sqrt{c} + b\sqrt{c} = (a + b)\sqrt{c}$ For example, $5\sqrt{2} + 3\sqrt{2} = (5 + 3)\sqrt{2} = 8\sqrt{2}$

$a\sqrt{c} - b\sqrt{c} = (a - b)\sqrt{c}$ For example, $6\sqrt{3} - 2\sqrt{3} = (6 - 2)\sqrt{3} = 4\sqrt{3}$

Collect terms where possible.

AQA Examiner's tip

Remember that you must know your squares up to 15^2. This means that you should also know the associated square roots.

Rationalising the denominator of a fraction means removing any square roots from the denominator.

If the denominator is $\sqrt{a}$ or a multiple of $\sqrt{a}$, then multiply the numerator and denominator by $\sqrt{a}$.

For example, $\frac{\sqrt{7}}{3\sqrt{2}} = \frac{\sqrt{7} \times \sqrt{2}}{3\sqrt{2} \times \sqrt{2}} = \frac{\sqrt{14}}{6}$ because $\sqrt{2} \times \sqrt{2} = \sqrt{4} = 2$

If the denominator is $b + \sqrt{c}$, then multiply the numerator and denominator by $b - \sqrt{c}$.

If the denominator is $b - \sqrt{c}$, then multiply the numerator and denominator by $b + \sqrt{c}$.

Link

Unit 2 Learn 4.2 gives more examples of rationalising denominators.

Aim higher

For Grade A and A*, you need to be able to rationalise the denominators of surds.

Example Factors and multiples Unit 2

C

a **i** Find all the common factors of 20, 28 and 36.

ii What is the highest common factor of 20, 28 and 36?

b Alice, Laura and Chloe all swim regularly at the same swimming pool.

Alice swims every 5 days, Laura swims every 6 days and Chloe swims every 10 days.
They all swam today. After how many days do they all swim again on the same day?

Solution

a **i** The factors of 20 are **1**, **2**, **4**, 5, 10, 20.
The factors of 28 are **1**, **2**, **4**, 7, 14, 28.
The factors of 36 are **1**, **2**, 3, **4**, 6, 9, 12, 18, 36.
The common factors of 20, 28 and 36 are 1, 2 and 4.

ii The highest common factor of 20, 28 and 36 is **4**.

b Alice swims again after 5 days, 10 days, 15 days, 20 days, 25 days, **30 days**, … These are the multiples of 5.
Laura swims again after 6 days, 12 days, 18 days, 24 days, **30 days**, …
Chloe swims again after 10 days, 20 days, **30 days**, …
They all swim again on the same day after 30 days. 30 is the Least Common Multiple of 5, 6 and 10.

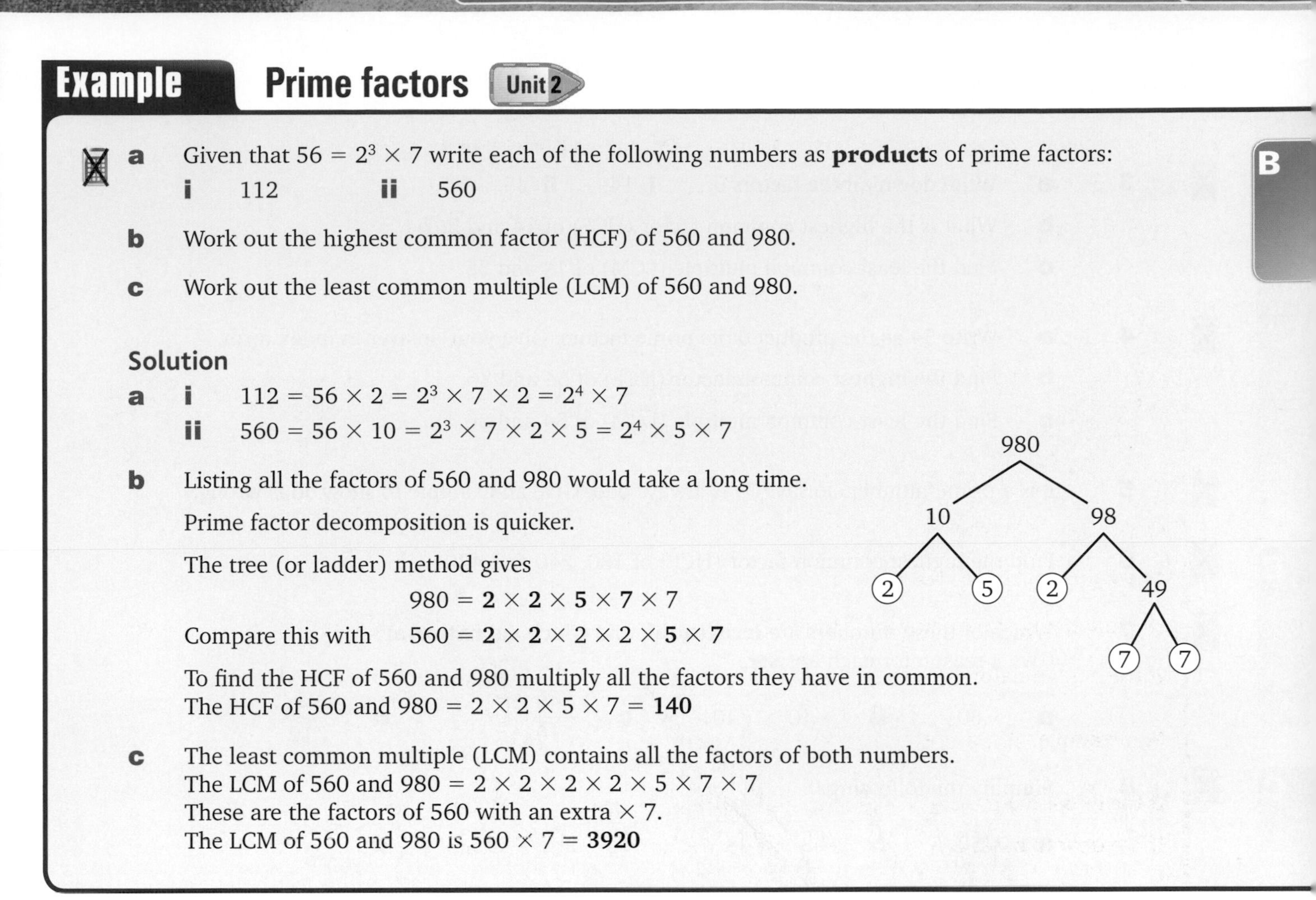

Example Prime factors Unit 2

B

a Given that $56 = 2^3 \times 7$ write each of the following numbers as **product**s of prime factors:

i 112 **ii** 560

b Work out the highest common factor (HCF) of 560 and 980.

c Work out the least common multiple (LCM) of 560 and 980.

Solution

a **i** $112 = 56 \times 2 = 2^3 \times 7 \times 2 = 2^4 \times 7$

ii $560 = 56 \times 10 = 2^3 \times 7 \times 2 \times 5 = 2^4 \times 5 \times 7$

b Listing all the factors of 560 and 980 would take a long time.

Prime factor decomposition is quicker.

The tree (or ladder) method gives

$980 = \mathbf{2} \times \mathbf{2} \times \mathbf{5} \times \mathbf{7} \times 7$

Compare this with $560 = \mathbf{2} \times \mathbf{2} \times 2 \times 2 \times \mathbf{5} \times \mathbf{7}$

To find the HCF of 560 and 980 multiply all the factors they have in common.
The HCF of 560 and 980 $= 2 \times 2 \times 5 \times 7 =$ **140**

c The least common multiple (LCM) contains all the factors of both numbers.
The LCM of 560 and 980 $= 2 \times 2 \times 2 \times 2 \times 5 \times 7 \times 7$
These are the factors of 560 with an extra $\times 7$.
The LCM of 560 and 980 is $560 \times 7 =$ **3920**

Example Surds Unit 2

A A*

Work out k where $\dfrac{3k + 1}{5 + \sqrt{3}} = 5 - \sqrt{3}$

Solution

Multiplying both sides of the equation by $5 + \sqrt{3}$ gives

$3k + 1 = (5 + \sqrt{3})(5 - \sqrt{3})$

$3k + 1 = 25 - 5\sqrt{3} + 5\sqrt{3} - 3$ because $\sqrt{3} \times \sqrt{3} = 3$

$3k + 1 = 22$

$3k = 21$ and so $k = 7$

Aim higher

You should be able to solve problems involving surds for a Grade A/A*.

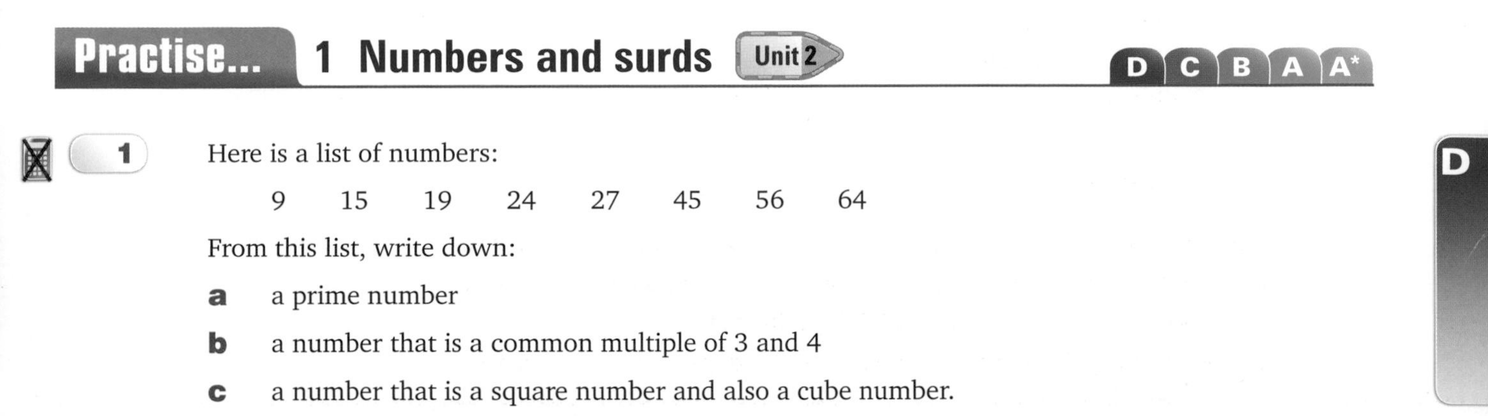

Practise... 1 Numbers and surds Unit 2

D C B A A*

D

1 Here is a list of numbers:

9 15 19 24 27 45 56 64

From this list, write down:

a a prime number

b a number that is a common multiple of 3 and 4

c a number that is a square number and also a cube number.

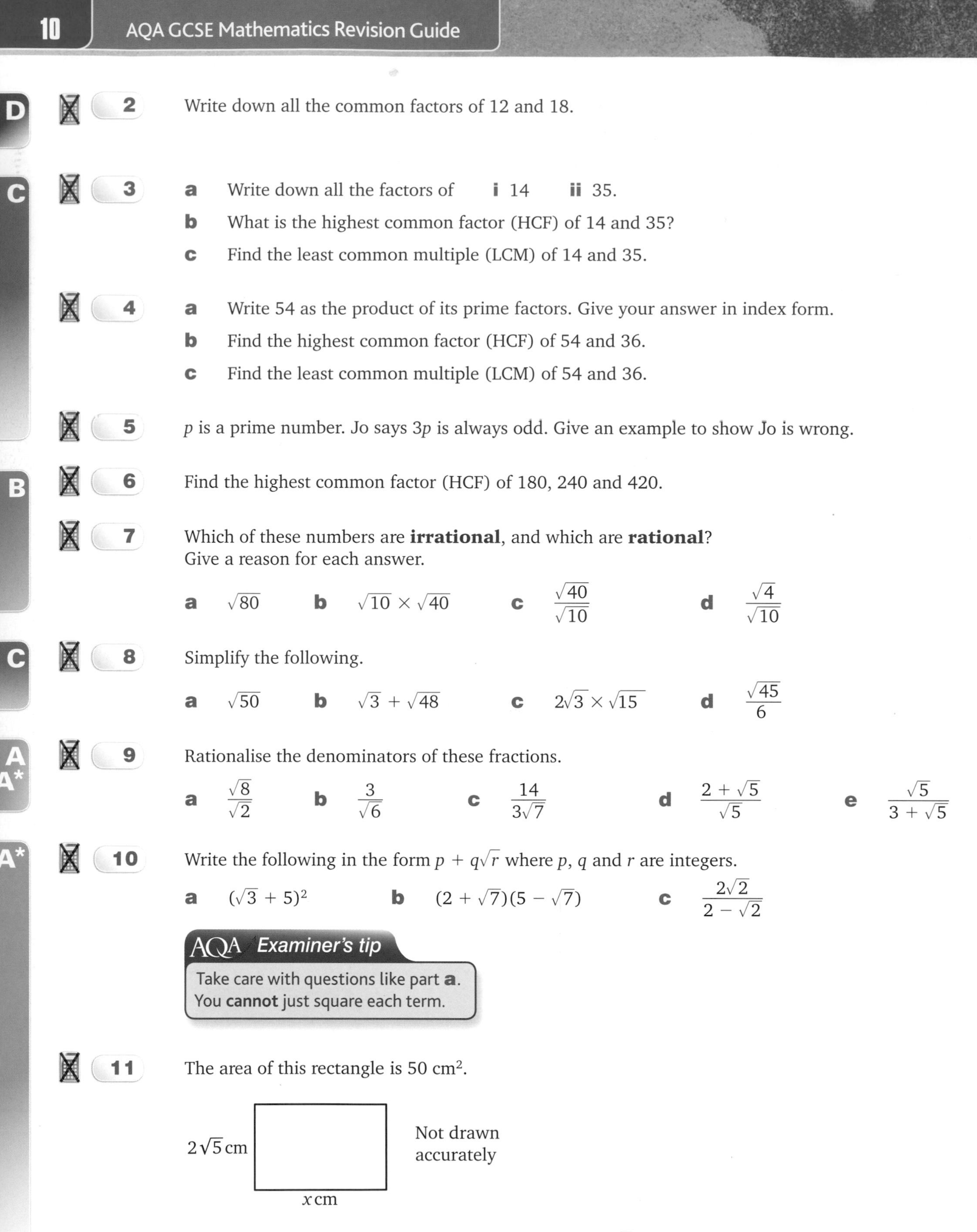

D

2 Write down all the common factors of 12 and 18.

C

3
- **a** Write down all the factors of **i** 14 **ii** 35.
- **b** What is the highest common factor (HCF) of 14 and 35?
- **c** Find the least common multiple (LCM) of 14 and 35.

4
- **a** Write 54 as the product of its prime factors. Give your answer in index form.
- **b** Find the highest common factor (HCF) of 54 and 36.
- **c** Find the least common multiple (LCM) of 54 and 36.

5 p is a prime number. Jo says $3p$ is always odd. Give an example to show Jo is wrong.

B

6 Find the highest common factor (HCF) of 180, 240 and 420.

7 Which of these numbers are **irrational**, and which are **rational**?
Give a reason for each answer.

a $\sqrt{80}$ **b** $\sqrt{10} \times \sqrt{40}$ **c** $\dfrac{\sqrt{40}}{\sqrt{10}}$ **d** $\dfrac{\sqrt{4}}{\sqrt{10}}$

C

8 Simplify the following.

a $\sqrt{50}$ **b** $\sqrt{3} + \sqrt{48}$ **c** $2\sqrt{3} \times \sqrt{15}$ **d** $\dfrac{\sqrt{45}}{6}$

A
A*

9 Rationalise the denominators of these fractions.

a $\dfrac{\sqrt{8}}{\sqrt{2}}$ **b** $\dfrac{3}{\sqrt{6}}$ **c** $\dfrac{14}{3\sqrt{7}}$ **d** $\dfrac{2 + \sqrt{5}}{\sqrt{5}}$ **e** $\dfrac{\sqrt{5}}{3 + \sqrt{5}}$

A*

10 Write the following in the form $p + q\sqrt{r}$ where p, q and r are integers.

a $(\sqrt{3} + 5)^2$ **b** $(2 + \sqrt{7})(5 - \sqrt{7})$ **c** $\dfrac{2\sqrt{2}}{2 - \sqrt{2}}$

AQA Examiner's tip

Take care with questions like part **a**.
You **cannot** just square each term.

11 The area of this rectangle is 50 cm².

Find the value of x, writing your answer in the form $a\sqrt{b}$ where a and b are integers.

12 Kate makes writing sets to sell at the school fair.

Each set contains a notepad, pencil and pencil sharpener.

She buys the notepads, pencils and pencil sharpeners in packs.

There are 10 pencils in each pack.

There are 6 notepads in each pack.

There are 4 pencil sharpeners in each pack.

She needs **exactly** the same number of notepads, pencils and pencil sharpeners.

What is the smallest number of each pack she must buy? You **must** show all your working.

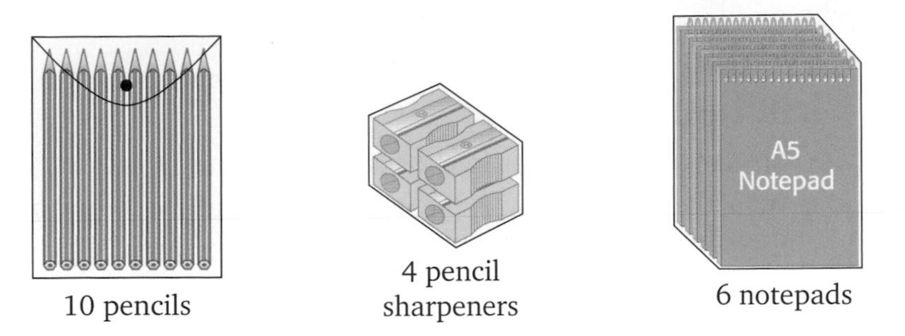

10 pencils 4 pencil sharpeners 6 notepads

13 Find two prime numbers between 50 and 70 whose difference is also a prime number.

14 You are given that $p = \sqrt{2} + 6$ and $q = \sqrt{2} - 6$

Work out k where $k(p + q) = p - q$. Give your answer as a surd in its simplest form.

15 A triangle has sides of length 1 cm, $(\sqrt{2} + \sqrt{6})$ cm and $(2 + \sqrt{3})$ cm.

Is this a right-angled triangle? Explain your answer.

Hint

The triangle has a right angle if $a^2 + b^2 = c^2$ (Pythagoras' theorem)

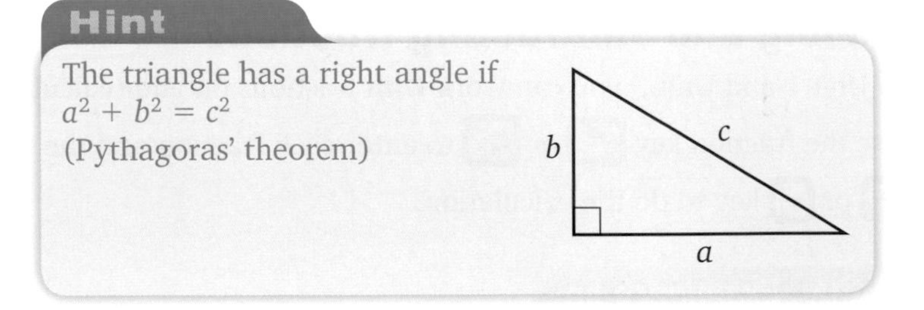

Number

2 Fractions, decimals and rounding

Unit 1

1 Fractions and decimals

Unit 2

1 Fractions and decimals

Unit 3

1 Fractions and decimals

Key terms

Write down definitions for the following words. Check your answers in the glossary of your Student Book.

decimal places
denominator
equivalent fractions
improper fraction
integer
lower bound
mixed number
numerator
recurring decimal
reciprocal
round
significant figures
terminating decimal
upper bound

Revise... Key points

Writing one quantity as a fraction of another

- Change both quantities into the same units first (if necessary).
- Write the first quantity as the **numerator** and the second as the **denominator** of a fraction.
- Simplify the fraction.

In Units 1 and 3, you can use your calculator to simplify the fraction, but in Unit 2 you will need to do this by dividing the numerator and denominator by the same numbers until you find the simplest **equivalent fraction**.

For example, to write 45 minutes as a fraction of $1\frac{1}{4}$ hours:

$1\frac{1}{4}$ hours = 75 minutes $\quad \frac{45}{75} = \frac{3}{5}$ (÷ 15) $\quad$ or $\quad \frac{45}{75} = \frac{9}{15} = \frac{3}{5}$ (÷ 5, ÷ 3)

45 minutes is $\frac{3}{5}$ of $1\frac{1}{4}$ hours.

Adding and subtracting fractions

All Units 1 2 3

In Unit 1 and Unit 3 you can work with fractions on your calculator.

Use the fraction key [fraction key] or [$a\frac{b}{c}$] to enter each fraction and the normal [+] or [−] key to do the calculation.

AQA Examiner's tip

Make sure that you know how to enter fractions and mixed numbers on your calculator for Units 1 and 3.

To add or subtract fractions without a calculator (Unit 2), write them with the same denominator.

Link

Unit 2 Learn 3.1 explains how to add and subtract fractions without a calculator.

To add or subtract mixed numbers without a calculator (Unit 2), first change them to **improper** (top-heavy) **fractions**.

Bump up your grade

You should be able to add and subtract mixed numbers for a Grade C.

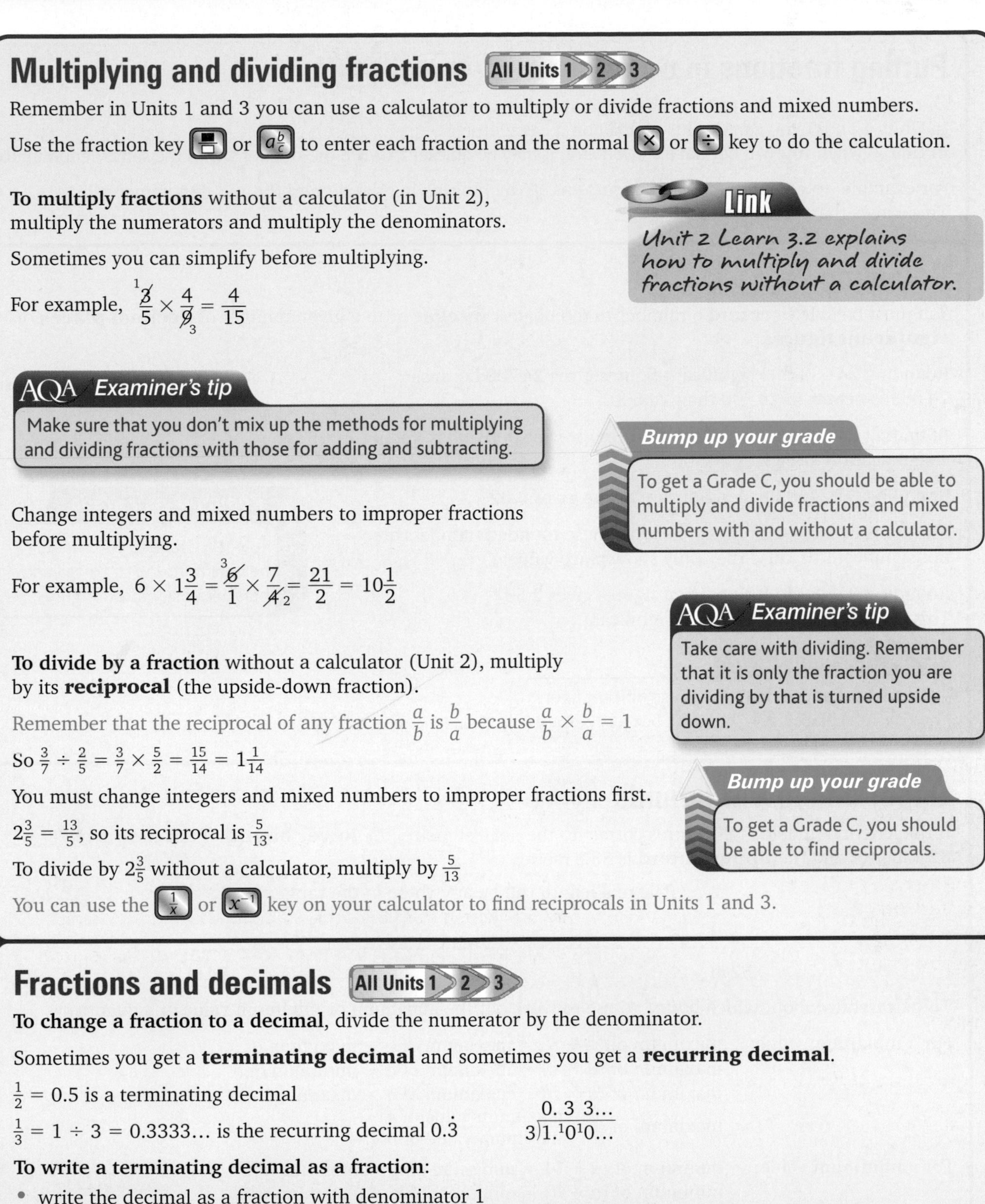

Multiplying and dividing fractions

All Units 1 2 3

Remember in Units 1 and 3 you can use a calculator to multiply or divide fractions and mixed numbers.

Use the fraction key or $a\frac{b}{c}$ to enter each fraction and the normal $\times$ or $\div$ key to do the calculation.

To multiply fractions without a calculator (in Unit 2), multiply the numerators and multiply the denominators.

Sometimes you can simplify before multiplying.

For example, $\frac{\not{3}^{1}}{5} \times \frac{4}{\not{9}_{3}} = \frac{4}{15}$

Link

Unit 2 Learn 3.2 explains how to multiply and divide fractions without a calculator.

AQA Examiner's tip

Make sure that you don't mix up the methods for multiplying and dividing fractions with those for adding and subtracting.

Bump up your grade

To get a Grade C, you should be able to multiply and divide fractions and mixed numbers with and without a calculator.

Change integers and mixed numbers to improper fractions before multiplying.

For example, $6 \times 1\frac{3}{4} = \frac{\not{6}^{3}}{1} \times \frac{7}{\not{4}_{2}} = \frac{21}{2} = 10\frac{1}{2}$

To divide by a fraction without a calculator (Unit 2), multiply by its **reciprocal** (the upside-down fraction).

Remember that the reciprocal of any fraction $\frac{a}{b}$ is $\frac{b}{a}$ because $\frac{a}{b} \times \frac{b}{a} = 1$

So $\frac{3}{7} \div \frac{2}{5} = \frac{3}{7} \times \frac{5}{2} = \frac{15}{14} = 1\frac{1}{14}$

You must change integers and mixed numbers to improper fractions first.

$2\frac{3}{5} = \frac{13}{5}$, so its reciprocal is $\frac{5}{13}$.

To divide by $2\frac{3}{5}$ without a calculator, multiply by $\frac{5}{13}$

You can use the $\frac{1}{x}$ or x^{-1} key on your calculator to find reciprocals in Units 1 and 3.

AQA Examiner's tip

Take care with dividing. Remember that it is only the fraction you are dividing by that is turned upside down.

Bump up your grade

To get a Grade C, you should be able to find reciprocals.

Fractions and decimals

All Units 1 2 3

To change a fraction to a decimal, divide the numerator by the denominator.

Sometimes you get a **terminating decimal** and sometimes you get a **recurring decimal**.

$\frac{1}{2} = 0.5$ is a terminating decimal

$\frac{1}{3} = 1 \div 3 = 0.3333...$ is the recurring decimal $0.\dot{3}$

$$3\overline{)1.{}^{1}0{}^{1}0...}\quad = 0.3\,3...$$

To write a terminating decimal as a fraction:

- write the decimal as a fraction with denominator 1
- multiply the numerator and denominator by 10 or 100 or 1000 ... to change it to an equivalent fraction with a whole number numerator
- simplify the fraction if possible.

For example: $0.36 = \frac{0.36}{1} = \frac{36}{100} = \frac{9}{25}$

To write a recurring decimal as a fraction (Unit 2 only):

- multiply it by 10 or 100 or 1000 ... so that the recurring part of the decimal stays the same
- subtract the original decimal – this eliminates the recurring part
- find the fraction by dividing, then simplifying if possible.

For example, writing the recurring decimal $0.\dot{1}\dot{5}$ as x.

$x = 0.15151515...$

and $100x = 15.151515...$

Subtracting gives $99x = 15$

so $x = \frac{15}{99} = \frac{5}{33}$

Putting fractions in order of size

All Units 1 2 3

One way is to write the fractions as decimals or percentages.
In Units 1 and 3 you can do this quickly on a calculator.
In Unit 2, when you will not have a calculator, it may be quicker to write the fractions with the same denominator.

For example, to compare $\frac{3}{5}$ with $\frac{11}{20}$, write $\frac{3}{5}$ as $\frac{12}{20}$ (by multiplying the numerator and denominator by 4).
This shows that $\frac{3}{5}$ is bigger.

Rounding

All Units 1 2 3

You must be able to **round** a number to the nearest **integer** or to a given number of **decimal places** or **significant figures**.

Rounding 24 683 to 3 significant figures gives 24 700 because 24 683 is nearer to 24 700 than 24 600.

Rounding 24 683 to 2 significant figures gives 25 000 and rounding 24 683 to 1 significant figure gives 20 000.

Rounding 0.024683 to 1 significant figure gives 0.02

The zeros at the beginning and end of these rounded numbers are not significant because they only show place value.

Rounding 2.4983 to 3 significant figures gives 2.50
This time the zero at the end is significant.

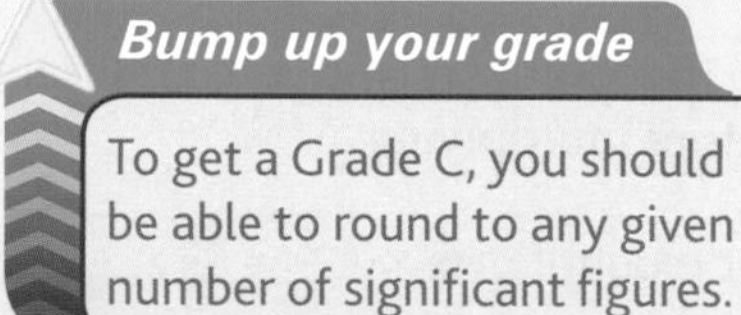

AQA Examiner's tip

Make sure that you don't mix up significant figures and decimal places. 2.47 has 3 s.f. but just 2 d.p.

Upper and lower bounds

Unit 1

If the length of a field is 83 metres correct to the nearest metre, the **lower bound** of this measurement is 82.5 metres and the **upper bound** is 83.5 metres.

The real length can be anywhere in the range:
$82.5 \leqslant \text{length} < 83.5$

82 m 82.5 m 83 m 83.5 m 84 m

Think carefully about which bound to use to work out the maximum or minimum value of a quantity.

For a **maximum** value:

maximum of $(a + b)$ = maximum of a + maximum of b
maximum of $(a - b)$ = maximum of a − minimum of b
maximum of $(a \times b)$ = maximum of a × maximum of b
maximum of $\frac{a}{b} = \frac{\text{maximum value of } a}{\text{minimum value of } b}$

For a **minimum** value:

minimum of $(a + b)$ = minimum of a + minimum of b
minimum of $(a - b)$ = minimum of a − maximum of b
minimum of $(a \times b)$ = minimum of a × minimum of b
minimum of $\frac{a}{b} = \frac{\text{minimum value of } a}{\text{maximum value of } b}$

Decimal calculations

All Units 1 2 3

Remember you are allowed to use a calculator in Units 1 and 3, but not in Unit 2.

To add (or **subtract**) decimals without a calculator, **line up the decimal points**.

To multiply decimal numbers without a calculator:

For example $1.2 \times 0.03 = 0.036$

- Remove the decimal points.
- Multiply the numbers as usual.
- Use an estimate to find out where to put the decimal point. Or count the total number of decimal places in the original numbers as the answer should have the same number of decimal places.

To divide decimal numbers without a calculator:

For example $1.2 \div 0.03 = \frac{1.2}{0.03} = \frac{120}{3} = 40$

- Write the division as a fraction.
- Multiply the numerator and denominator so that the denominator is a whole number.
  If the denominator has 1 decimal place, you will need to multiply by 10.
  If the denominator has 2 decimal places, you will need to multiply by 100.
- Divide the numerator by the denominator to find the answer.

Link

Unit 2 Learn 3.7 includes a range of decimal calculations done without a calculator.

Bump up your grade

For a Grade C, you should be able to divide by a decimal without a calculator.

AQA Examiner's tip

Remember that multiplying by a number less than 1 gives an answer that is smaller than the original. Dividing by a number less than 1 gives an answer that is bigger than the original.

Example Calculating with mixed numbers Unit 2

C

A dressmaker makes trouser suits for a children's clothes shop.
Each suit consists of a jacket and trousers.

The dressmaker uses $1\frac{5}{6}$ metres of fabric for each jacket and $1\frac{1}{2}$ metres for each pair of trousers.

How many suits can she make from a 25-metre roll of fabric?

Solution

The total amount of fabric for one suit$= 1\frac{5}{6} + 1\frac{1}{2} = \frac{11}{6} + \frac{3}{2} = \frac{11}{6} + \frac{9}{6} = \frac{20}{6} = \frac{10}{3}$

Number of suits she can make from 25 metres $= 25 \div \frac{10}{3} = \frac{{}^{5}\cancel{25}}{1} \times \frac{3}{\cancel{10}_{2}} = \frac{15}{2} = 7\frac{1}{2}$

She can make 7 trouser suits.

Make sure that you can also do calculations like these on your calculator for Units 1 and 3.

Example Working with fractions and decimals Unit 2

C

A water tank is $\frac{4}{5}$ full. After 7.5 litres of water are used from the tank it is $\frac{3}{4}$ full.

How much does the tank hold when it is full?

Solution

The fraction that was used $= \frac{4}{5} - \frac{3}{4} = \frac{16}{20} - \frac{15}{20} = \frac{1}{20}$

$\frac{1}{20}$ of a tank = 7.5 litres, so a full tank holds $7.5 \times 20 = 150$ litres.

Example Recurring decimals Unit 2

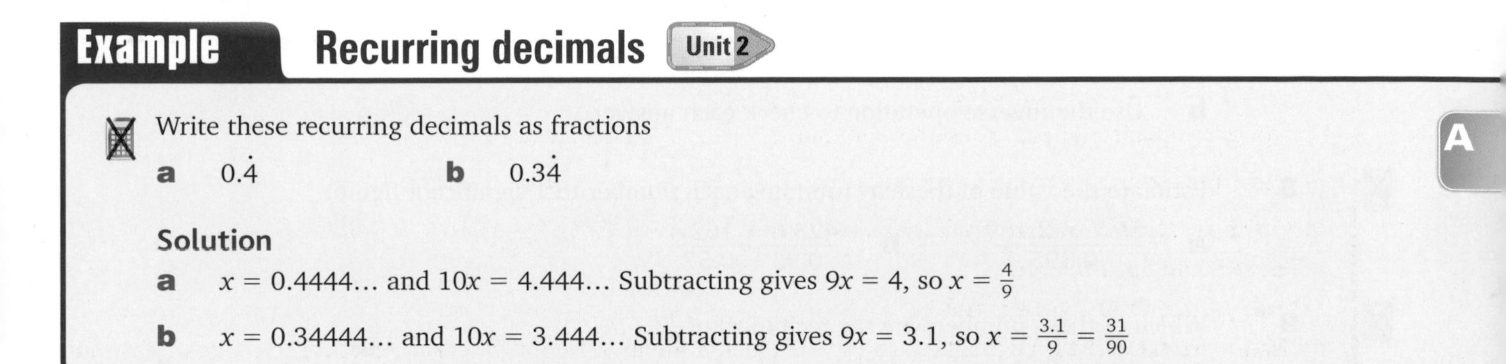

A

Write these recurring decimals as fractions

a $0.\dot{4}$ **b** $0.3\dot{4}$

Solution

a $x = 0.4444\ldots$ and $10x = 4.444\ldots$ Subtracting gives $9x = 4$, so $x = \frac{4}{9}$

b $x = 0.34444\ldots$ and $10x = 3.444\ldots$ Subtracting gives $9x = 3.1$, so $x = \frac{3.1}{9} = \frac{31}{90}$

Example Upper and lower bounds Unit 1

A*

A shelf can support a weight of 20 kilograms, correct to the nearest kilogram.
A teacher wants to put some identical books on the shelf.
Each book weighs 1.2 kilograms, correct to 2 significant figures.
What is the maximum number of books that the teacher should put on the shelf? Explain your answer.

Solution

To be safe, the teacher should assume that the shelf can only support 19.5 kg (the lower bound of the given value) and that the weight of each book is 1.25 kg (the upper bound of the given value).

Maximum number of books $= \frac{19.5}{1.25} = 15$

$19.5 \div 1.25 = 15.6$, but this is rounded **down** to give a **whole** number of books.

AQA Examiner's tip

Think very carefully about which bounds to use – you do not get any marks for doing the calculation with the original numbers or the wrong bounds.

Aim higher

You will need to be able to solve problems involving upper and lower bounds for a Grade A/A*.

Practise... Fractions, decimals and rounding

All Units 1 2 3

D C B A A*

D

1 Use your calculator to work out $\frac{2.46 + 5.89}{3.92 - 1.75}$

- **a** Write down all the figures in your calculator display.
- **b** Round your answer to part **a** **i** to 1 decimal place **ii** to 1 significant figure.

2 Write the following fractions as decimals. **a** $\frac{13}{20}$ **b** $\frac{7}{8}$ **c** $\frac{2}{3}$ **d** $\frac{7}{9}$ **e** $1\frac{1}{11}$

3 24 dancers out of 60 pass an audition.

- **a** What fraction of the dancers pass the audition?
- **b** What fraction of the dancers fail the audition?

Give each fraction in its simplest form.

4 Which of the following fractions is nearest to $\frac{3}{4}$? Show how you decide.

$\frac{5}{8}$ $\frac{7}{10}$ $\frac{3}{5}$

5 Work these out.

a $\frac{5}{6} + \frac{8}{9}$ **b** $\frac{5}{8} - \frac{2}{5}$ **c** $\frac{6}{7} \times \frac{3}{4}$ **d** $\frac{9}{10} \div \frac{3}{5}$

6 Write these decimals as fractions.
Give each answer in its simplest form.

a 0.85 **c** 0.375 **e** 0.0625

b 0.56 **d** 0.016

AQA Examiner's tip

Make sure that you can do all these calculations on a calculator too for Units 1 and 3.

D C

7 **a** Work these out. **i** 2.4×0.3 **ii** $7.56 \div 0.6$

b Use the inverse operation to check each answer.

C

8 Estimate the value of these by rounding each number to 1 significant figure.

a $\frac{59.7 \times 2.189}{0.395}$ **b** $\frac{423.6 + 162.4}{0.93 - 0.57}$

9 Which of these numbers are the reciprocal of $\frac{4}{5}$?

a 0.45 **b** $\frac{5}{4}$ **c** $1\frac{1}{5}$ **d** 0.8 **e** 1.25 **f** 1.2 **g** $1\frac{1}{4}$

C

10 Work these out.

a $2\frac{3}{5} + 4\frac{1}{2}$ **b** $5\frac{2}{3} - 1\frac{5}{6}$ **c** $3\frac{1}{5} \times 2\frac{3}{4}$ **d** $2\frac{5}{8} \div 3\frac{1}{2}$ **e** $2\frac{8}{9} - \frac{5}{6} \times 1\frac{2}{3}$

11 A shop usually sells rugs for a third more than it pays for them.
In a sale it reduces the selling price by a third.
A shop assistant says the shop is making a loss on the rugs it sells in the sale.
Is the shop assistant right? Explain your answer.

B

12 **a** Express $\frac{7}{11}$ as a recurring decimal.

b Which of the following fractions are recurring decimals? Explain your answer.

$\frac{7}{8}$ $\frac{7}{9}$ $\frac{7}{20}$ $\frac{1}{7}$

A

13 Component A is 134 mm long and component B is 76 mm long.
Both of these lengths are given to the nearest millimetre.

a What is the maximum total length of these components?

b What is the maximum difference in length of these components?

A A*

14 Express each recurring decimal as a fraction in its simplest form.

a $0.\dot{8}$ **b** $0.\dot{4}\dot{5}$ **c** $0.0\dot{4}\dot{5}$ **d** $0.2\dot{6}$ **e** $0.\dot{2}1\dot{6}$

15 Sam uses $\frac{3}{4}$ of a tin of paint to paint a door.
Work out the **smallest** number of tins he needs to paint 6 identical doors.

16 The table gives the costs of hiring a chainsaw and safety equipment.

Item	First day	Each extra day	Weekend (Sat & Sun)	Week
Chainsaw	£44.90	£12.50	£54.90	£80.90
Chainsaw safety equipment	£12.10	£4.95	£8.25	£22.40

How much cheaper is it to hire the chainsaw and safety equipment at the weekend than for 2 days midweek?

17 Bill earns a basic wage of £7.24 per hour for a 37-hour week. He is paid 'time and a quarter' for any overtime. How many hours' overtime does Bill need to work to earn over £300 in a week?

18 Adam uses $\frac{2}{3}$ litre of milk a day.
He wants to buy enough milk to last for a week for the cheapest price.
The shop sells milk in 2-pint, 4-pint and 6-pint bottles as shown. What should Adam buy?
[Assume 1 litre = $1\frac{3}{4}$ pints]

MILK 2 pint 80p

MILK 4 pint £1.50

MILK 6 pint £2.10

19 A company sells toffees in bags of 30. Each toffee weighs 10 grams, correct to the nearest gram.
What should be advertised as the minimum contents by weight of each bag of toffees?

20 A crane can lift a load of 450 kg, correct to 2 significant figures.
How many blocks of stone, each of mass 16 kg correct to 2 significant figures, can it be sure of lifting?

21 Kieran says that $\frac{1}{15}$ is halfway between $\frac{1}{10}$ and $\frac{1}{20}$.
Is Kieran correct? Show how you decide.

22 Find a fraction between $\frac{3}{4}$ and $\frac{4}{5}$ that is a recurring decimal.

Number

3 Percentages

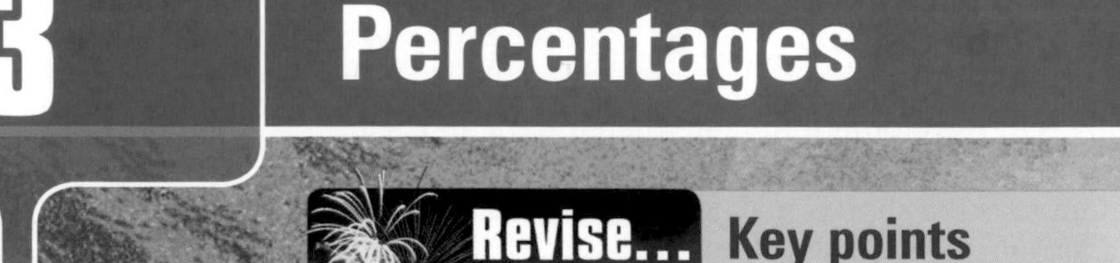

Key terms

Write down definitions for the following words. Check your answers in the glossary of your Student Book.

balance
deposit
depreciation
discount
interest
percentage
principal
rate
Value Added Tax (VAT)

Increasing or decreasing an amount by a percentage

Multiplier method

This is the most efficient method on a calculator (Units 1 and 3).

- Write the new quantity as a **percentage** of the original quantity.
- Convert this percentage to a **multiplier** (by dividing it by 100).
  For a 35% increase, the multiplier is $135 \div 100 = 1.35$.
  For a 35% decrease, the multiplier is $65 \div 100 = 0.65$
- Multiply by the original quantity.

This method is the best way to combine percentage changes and work out compound interest in Unit 1.

Unitary method

- **Divide** the quantity **by 100** (to find 1%).
- Then **multiply by the percentage** you need.
- For a percentage **increase**, **add** to the original amount.
  For a percentage **decrease**, **subtract** from the original amount.

Using links with 10%

This is often the easiest method when you don't have a calculator (Unit 2).

- Use links with 10% (and/or 1%) to find the increase (or decrease)
- For a percentage **increase**, **add** to the original amount. For a percentage **decrease**, **subtract** from the original amount.

AQA Examiner's tip

It is always a good idea to **check** your answers. Use a different method if you can.

Writing one quantity as a percentage of another

All Units 1 2 3

- Write both quantities in the same units.
- Divide the first quantity by the second to give a fraction or decimal.
- Write the fraction or decimal as a percentage.

Writing an increase and decrease as a percentage

- Find the increase or decrease.
- Divide the increase (or decrease) by the **original** amount to give a fraction or decimal.
- Write the fraction or decimal as a percentage.

$$\text{percentage increase (or decrease)} = \frac{\text{increase (or decrease)}}{\textbf{original}\text{ amount}} \times 100\%$$

You can also use this method to find a percentage profit or loss.

$$\text{percentage profit (or loss)} = \frac{\text{profit (or loss)}}{\textbf{original}\text{ (cost) price}} \times 100\%$$

AQA Examiner's tip

Remember that you must use the same units when writing one quantity as a percentage of another.

Bump up your grade

For a Grade C, you should be able to write an increase or decrease as a percentage.

Reverse percentage problems

Units 1 2

In a reverse percentage problem, you start with the final amount and work back to the original amount.

Multiplier method – this is the most efficient method on a calculator (Unit 1).

- Write the new amount as a percentage of the original amount.
- Convert this percentage to a **multiplier** (by dividing it by 100).
- **Divide** the final amount by the multiplier to find the original amount.

Unitary method – this is usually the easiest method to use without a calculator (Unit 2). Also useful on a calculator (Unit 1).

- Write the new amount as a percentage of the original amount.
- Divide the new amount by this percentage to **find 1%**.
- Multiply by 100 to find 100% of the original amount.

Example Increasing by a percentage

D

Sally earns £17 520 per year. She gets a pay rise of $2\frac{1}{2}$%.
How much does she now get per month?

AQA Examiner's tip

Write down what you are doing, so that you get marks for the method even if you make a mistake on your calculator.

Solution

Sally's increased pay = 102.5% of her previous pay
The multiplier = 102.5 ÷ 100 = 1.025
Sally's new pay per month = 1.025 × £17 520 ÷ 12
= £1496.50

You can do this on a calculator in a single calculation. You could use the unitary method to check.

Example Decreasing by a percentage

Unit 2

C

The usual cost of insuring Aram's car is £450. His insurance company gives him a **discount** of 20% for not claiming in previous years. The discounted price is reduced by another 15% if Aram agrees to pay the first £100 of any claim he makes.

How much does Aram pay?

Bump up your grade

For a Grade C, you need to be able to carry out more than one percentage change.

Solution

10% of £450 = £450 ÷ 10 = £45
20% of £450 = £45 × 2 = £90
The discounted price = £450 − £90 = £360

10% of £360 = £360 ÷ 10 = £36
5% of £360 = £36 ÷ 2 = £18
15% of £360 = £36 + £18 = £54
The amount Aram pays = £360 − £54 = £306

The multiplier method is more efficient when a calculator is available (Units 1 and 3):
The amount Aram pays = 85% of 80% of £450
= 0.85 × 0.80 × £450
= £306

Example Writing one quantity as a percentage of another

Units 1 3

D

762 students at a school are right-handed and 92 are left-handed.
What percentage of the students are left-handed? Give your answer correct to 1 decimal place.

Solution

The total number of students = 762 + 92 = 854

Percentage who are left-handed = $\frac{92}{854}$ × 100% = 92 ÷ 854 × 100%
= 10.7728… = 10.8% to 1 decimal place

You need to write the number of left-handed students as a percentage of the total number of students.

Example Percentage increase Unit 2

C

A shopkeeper buys notebooks in packs of 20 for £24 per pack.
She sells the notebooks for £1.80 each.
What is her percentage profit?

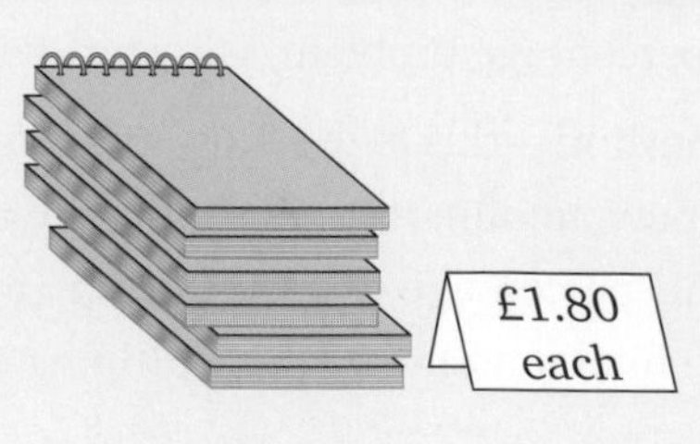

Solution

Cost price for 1 notebook = £24 ÷ 20 = £1.20
The profit on each notebook = £1.80 − £1.20 = 60 pence

Writing the profit as a fraction of the **original** (cost) price gives $\frac{60}{120} = \frac{1}{2}$. The percentage profit = 50%.

An alternative method is to work out the % profit on the pack of 20 notebooks.
The shopkeeper sells 20 notebooks for 20 × £1.80 = 2 × £18 = £36
The profit = £36 − £24 = £12
Writing this as a fraction of the original (cost) price gives $\frac{12}{24} = \frac{1}{2} = 50\%$

AQA Examiner's tip

Remember to use the same units and that the denominator should be the **original** amount.

Bump up your grade

You should be able to work out percentage increases and decreases for a Grade C.

Example Reverse percentage Unit 1

B

A charity can claim back the **VAT** it has spent on some equipment.
The equipment cost £259.14 including VAT at a rate of 5%.
How much can the charity claim back?

Solution

Using the unitary method:
105% of the cost of the equipment = £259.14
1% of the cost of the equipment = £259.14 ÷ 105 = £2.468
5% of the cost of the equipment (the VAT) = £2.468 × 5 = £12.34
so the charity can claim back £12.34.

Alternatively, using the multiplier method:
£259.14 ÷ 1.05 × 0.05 = £12.34 where 1.05 is the multiplier for the total cost and 0.05 is the multiplier for the VAT.

Example Compound interest Unit 1

B

Val invests a **principal** of £2000.
The account pays **interest** at a **rate** of 4.7% at the end of each year.

a Find the amount Val will have in the account after 5 years.

b Show that it will take over 15 years for Val's investment to double in value.

Solution

a The amount in the account at the end of each year is 104.7% of the amount at the beginning of the year.
So the multiplier is 1.047.
After 5 years, the amount in the account = £2000 × 1.047^5 = £2516.31

b After 15 years, the amount = £2000 × 1.047^{15} = £3983.18…
The investment of £2000 has not quite doubled in value.
So it takes over 15 years for the amount to double in value.

Repeated percentage increases lead to exponential growth.
Compound interest is one example of this.

Practise... Percentages

D C B A A*

D

1 **a** Use multipliers to:

i increase £135 by 6% **ii** reduce £135 by 6%

b Check your answers using a different method.

D C

2 Mr Marks gives his class a test.
The table shows their marks out of 25.

Marks	Number of students
0–4	1
5–9	5
10–14	10
15–19	8
20–25	6

Mr Marks says the pass mark is 40%.

a How many students pass the test?

b What percentage of the class fail the test?

C

3 In a sale the price of a tent goes down by £11 to £44.
What is the percentage reduction?

Hint
Make sure that you can also do these with a calculator (for Units 1 and 3)

4 The price of a magazine goes up from 96 pence to £1.20.
What is the percentage increase?

5 This year Jack's pay has risen from £17 500 to £18 200.
Jill's pay has risen from £24 000 to £24 840.
Whose pay rose by the greater percentage? You **must** show your working.

6 Three shops advertise the same mountain bike.

Shop A	Shop B	Shop C
Pay £150 **deposit** plus the **balance** in 6 equal monthly payments of £50	20% off usual price of £560	£400 plus VAT at $17\frac{1}{2}$%

Shelley wants to buy the cheapest one. Which shop should she go to?

7 A shop sells boxes of chocolates. The manager buys 60 boxes at £5 each.
The shop sells two-thirds of these boxes at a profit of 30%
In a sale, the shop sells the rest of the boxes of chocolates at half the usual selling price.
Work out the overall percentage profit.

B

8 After a reduction of 15%, a camera is priced at £136.
What was the price before the reduction?

9 A shop usually sells goods for 40% more than it paid for them.
In a sale, the shop reduces its prices by 20%
Work out the percentage profit that the shop makes on the goods it sells in the sale.

10 Julie invests £6500 at a fixed interest rate of 4% per year.

a How much is in Julie's account after interest is added at the end of the fifth year?

b Show that it will take 11 years for the amount in the account to grow to £10 000.

11 A delivery van is worth £27 500 when new and its value **depreciates** by 12% each year.

a Work out how much the van is worth when it is three years old.

b Show that the value of the van is halved within six years.

c Sketch a graph to show how the value of the van changes with time.

Number

4 Ratio and proportion

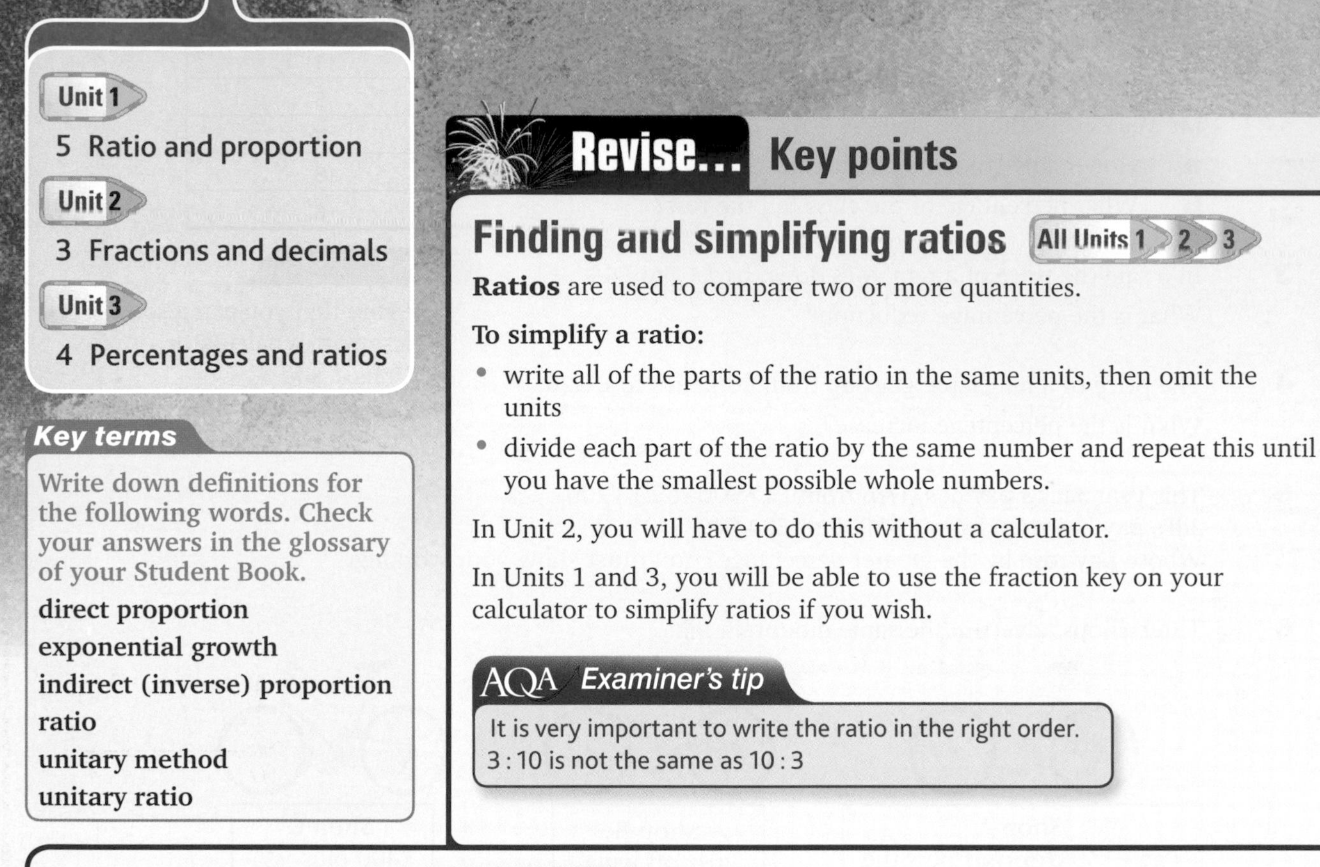

Key terms

Write down definitions for the following words. Check your answers in the glossary of your Student Book.

direct proportion
exponential growth
indirect (inverse) proportion
ratio
unitary method
unitary ratio

Revise... Key points

Finding and simplifying ratios (All Units 1 2 3)

Ratios are used to compare two or more quantities.

To simplify a ratio:

- write all of the parts of the ratio in the same units, then omit the units
- divide each part of the ratio by the same number and repeat this until you have the smallest possible whole numbers.

In Unit 2, you will have to do this without a calculator.

In Units 1 and 3, you will be able to use the fraction key on your calculator to simplify ratios if you wish.

AQA *Examiner's tip*

It is very important to write the ratio in the right order. 3 : 10 is not the same as 10 : 3

Links with fractions (All Units 1 2 3)

There are a variety of links between a ratio and fractions.

For example, when a brother and sister share some money in the ratio 2 : 3, this means for every £2 the brother gets, the sister gets £3.

The brother gets $\frac{2}{3}$ of what the sister gets. The sister gets $\frac{3}{2} = 1\frac{1}{2}$ times what the brother gets.

The brother gets $\frac{2}{5}$ of the total amount. The sister gets $\frac{3}{5}$ of the total amount.

Unitary ratios (All Units 1 2 3)

Sometimes it is useful to write a ratio in the form $1 : n$ or $n : 1$ where n may not be a whole number.

For example, in a school where there are 740 students and 50 teachers, the student : teacher ratio is 740 : 50
Dividing both sides by 50 gives 14.8 : 1.

This **unitary ratio** tells us that there are 14.8 students for every teacher at this school.
It is a useful ratio to use when comparing schools with different numbers of students and teachers.

In Units 1 and 3, you will be able to use a calculator.

In Unit 2, you need to be able to work it out without a calculator. You could do this by dividing both sides of the ratio by 10, then 5.

Unitary ratios are also used on plans, maps and models. For example, the ratio 1 : 100 is often used on plans, the ratio 1 : 1250 is used in town planning and the ratios 1 : 25 000 and 1 : 50 000 are used on Ordnance Survey maps.

To find the real distance between two places shown on a 1 : 50 000 map, multiply the distance on the map by 50 000.

Ratio and proportion: the unitary method

All Units 1 2 3

The **unitary method** is useful in many situations involving ratio and **proportion**. It is based on finding the amount or cost of **one** unit.

For example, a recipe says you need 450 g of tomatoes to make 6 portions of tomato soup, but you want 8 portions.

First work out the amount of tomatoes for **1 portion** of soup = 450 g ÷ 6 = 75 g

Then the amount of tomatoes for 8 portions = 75 g × 8 = 600 g

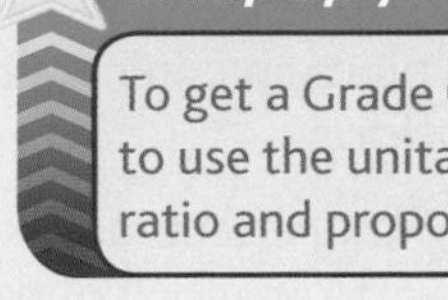

Bump up your grade

To get a Grade C, you should be able to use the unitary method to solve ratio and proportion problems.

One of the examples later shows how to use the unitary method to solve best-buy problems.

Dividing quantities in a given ratio

Unit 2

To divide something in a given ratio

- Add the numbers in the ratio to find the total number of parts.
- Divide the quantity by the total number of parts to find the amount for 1 part.
- Multiply by each number in the ratio to find the quantities required.

AQA Examiner's tip

At the end, check by adding. The total should be the same as the quantity you started with.

Direct and indirect (inverse) proportion

Unit 3

Direct proportion

y* is directly proportional to *x is written as $y \propto x$.
This means that $y = kx$ where k is the constant of proportionality.
The graph of y against x is a **straight line through the origin, (0, 0)** with **gradient k**.

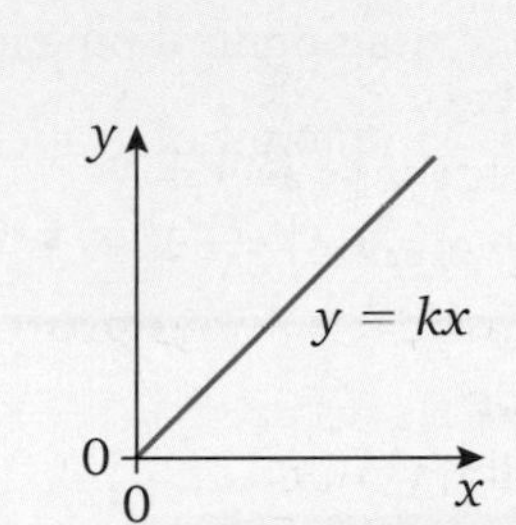

Unit 2 Chapter 6 $y = mx + c$ represents a straight line with gradient m and intercept $(0, c)$.

y is directly proportional to x^2, written $y \propto x^2$, means $y = kx^2$

y is directly proportional to x^3, written $y \propto x^3$, means $y = kx^3$

y is directly proportional to $\sqrt{x}$, written $y \propto \sqrt{x}$, means $y = k\sqrt{x}$

k is the constant of proportionality

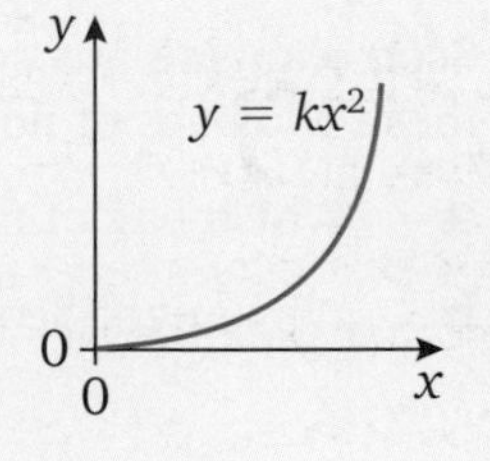

Indirect (inverse) proportion

y is **indirectly proportional** to x, written $y \propto \frac{1}{x}$, means $y = \frac{k}{x}$

y is indirectly proportional to x^2, written $y \propto \frac{1}{x^2}$, means $y = \frac{k}{x^2}$

y is indirectly proportional to $\sqrt{x}$, written $y \propto \frac{1}{\sqrt{x}}$, means $y = \frac{k}{\sqrt{x}}$

and so on.

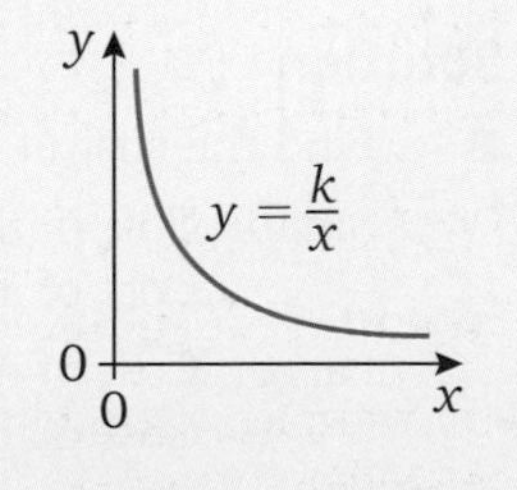

Exponential growth

Repeated proportional change leads to exponential growth or decay.

The function $y = Ak^{mx}$ (where A and m are positive constants) gives exponential growth when $k > 1$ and exponential decay when $0 < k < 1$.

For example, if £2000 earns interest at a rate of 4.7% per year, the amount after x years is £2000 × 1.047^x In this case A is 2000, k is 1.047 and m is 1.
The more money there is in the account, the faster it grows.

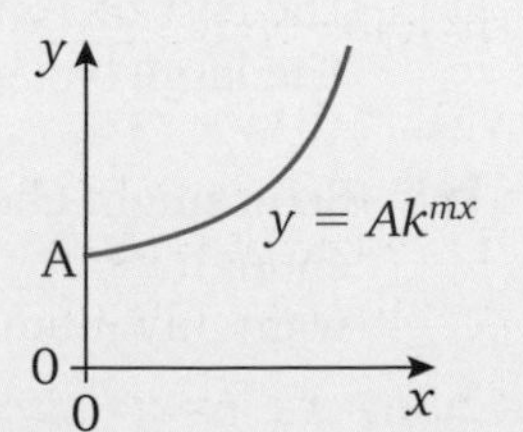

Aim higher

For a Grade A, you should be able to solve problems on direct and indirect proportionality and exponential growth.

Example Simplifying ratios and dividing in a ratio

Unit 2

C

Mrs Perfect uses this recipe to make a breakfast drink.

a Write the ratio of orange juice : pineapple juice : grapefruit juice in its simplest form.

b One morning Mrs Perfect makes $2\frac{1}{2}$ litres of breakfast drink. How much of each ingredient does she use?

Breakfast Fruit Juice

1 litre of orange juice
600 ml of pineapple juice
400 ml of grapefruit juice

Solution

a The ratio of orange juice, pineapple juice and grapefruit juice is 1000 ml : 600 ml : 400 ml = 5 : 3 : 2

b $2\frac{1}{2}$ litres = 2500 ml

The total number of parts in the ratio is 5 + 3 + 2 = 10

This means that $\frac{5}{10}$ of the drink is orange juice, $\frac{3}{10}$ is pineapple juice and $\frac{2}{10}$ is grapefruit juice.

$\frac{1}{10}$ of 2500 ml = 2500 ml ÷ 10 = 250 ml

amount of orange juice = $\frac{5}{10}$ of 2500 ml = 250 ml × 5 = 1250 ml

amount of pineapple juice = $\frac{3}{10}$ of 2500 ml = 250 ml × 3 = 750 ml

amount of grapefruit juice = $\frac{2}{10}$ of 2500 ml = 250 ml × 2 = 500 ml

To check this, work out 1250 ml + 750 ml + 500 ml = 2500 ml ✓

AQA Examiner's tip

Remember the units must be the same before you simplify the ratio.

AQA Examiner's tip

Check that the amounts add up to the total.

Example Using the unitary method to solve a best-buy problem

Units 1 3

C

Scott wants to buy some noodles. The supermarket sells these two bags of noodles.

a Which bag of noodles gives the best value for money?

b Why might Scott choose to buy the other bag?

Solution

a Find the cost of **1 gram** in each bag.

Small: Cost of 300 g = 189 pence
Cost of 1 g = 189 ÷ 300 = 0.63 pence

Large: Cost of 450 g = 249 pence
Cost of 1 g = 249 ÷ 450 = 0.5533… pence

The cost of 1 gram of noodles is less in the large bag. The large bag gives the best value for money.

b Scott might choose a small bag if he does not have enough money for the large bag or if he does not want any more than 300 g.

There are sometimes other ways to solve best-buy problems. In this case, you could find the cost of 50 g:

Small bag (as 300 g = 6 × 50 g):
cost of 50 g = 189 ÷ 6 = 31.5 pence.

Large bag (as 450 g = 9 × 50 g):
cost of 50 g = 249 ÷ 9 = 27.666… pence

This also shows that the large bag gives the best value for money.

Bump up your grade

To get a Grade C, you should be able to solve best buy problems.

Example Direct proportionality Unit 3

The distance, d kilometres, you can see to the horizon is directly proportional to the square root of your height above sea level, h metres. When you are 100 m above sea level, the distance to the horizon is 36 km.

Jason lives in a tower block by the sea. Work out:

a the distance to the horizon when Jason looks at it from a height of 25 metres

b how high Jason would have to be to see 25 km out to sea. (Give your answer to the nearest metre.)

Solution

$d \propto \sqrt{h}$ means $d = k\sqrt{h}$ where k is a constant.

Since $d = 36$ when $h = 100$, $36 = k\sqrt{100}$, so $36 = 10k$ and $k = 3.6$

The relationship between d and h is $d = 3.6\sqrt{h}$

a When $h = 25$, $d = 3.6\sqrt{25} = 3.6 \times 5 = 18$
When Jason looks at the horizon from a height of 25 metres, the distance to the horizon is 18 km.

Alternative method for **a** $d = k\sqrt{h}$ and when $h = 100$, $d = 36$

If h is divided by 4, then d is divided by $\sqrt{4}$. So when $h = 25$, $d = 36 \div 2 = 18$

b When $d = 25$, $25 = 3.6\sqrt{h}$ so $\sqrt{h} = \frac{25}{3.6} = 6.9\dot{4}$
$h = 6.9\dot{4}^2 = 48.225\ldots$ so to see 25 km out to sea,
Jason would have to be at a height of 48 metres (to the nearest metre).

Practise... Ratio and proportion All Units 1 2 3

D C B A A*

1 The ratio of girls to boys in a class is 5 : 4.

a What fraction of the class are girls?

b What fraction of the class are boys?

c There are 27 students in the class. How many of these are girls and how many are boys?

2 The angles of any pentagon add up to 540°.
The angles of one pentagon are in the ratio 2 : 3 : 4 : 5 : 6.
Find the size of the smallest angle.

3 The ratio of male to female workers in a factory is 7 : 13.
What percentage of the workers is female?

4 In a triathlon competitors swim 1500 m, cycle 40 km, then run 10 km.
Write the ratio of these distances in its simplest form.

AQA Examiner's tip
Make sure you can also do questions like these with a calculator.

5 In a pie chart about how people travel to work, an angle of 60° represents 75 people.

a How many people does an angle of 40° represent?

b What angle represents 120 people?

6 A pet shop owner has four budgies for sale.
He knows that one bag of bird food will last these budgies for six days.
A customer buys one of the budgies.
How long will a bag of bird food last now?

Hint
How long would the bag of bird food last **one** budgie?

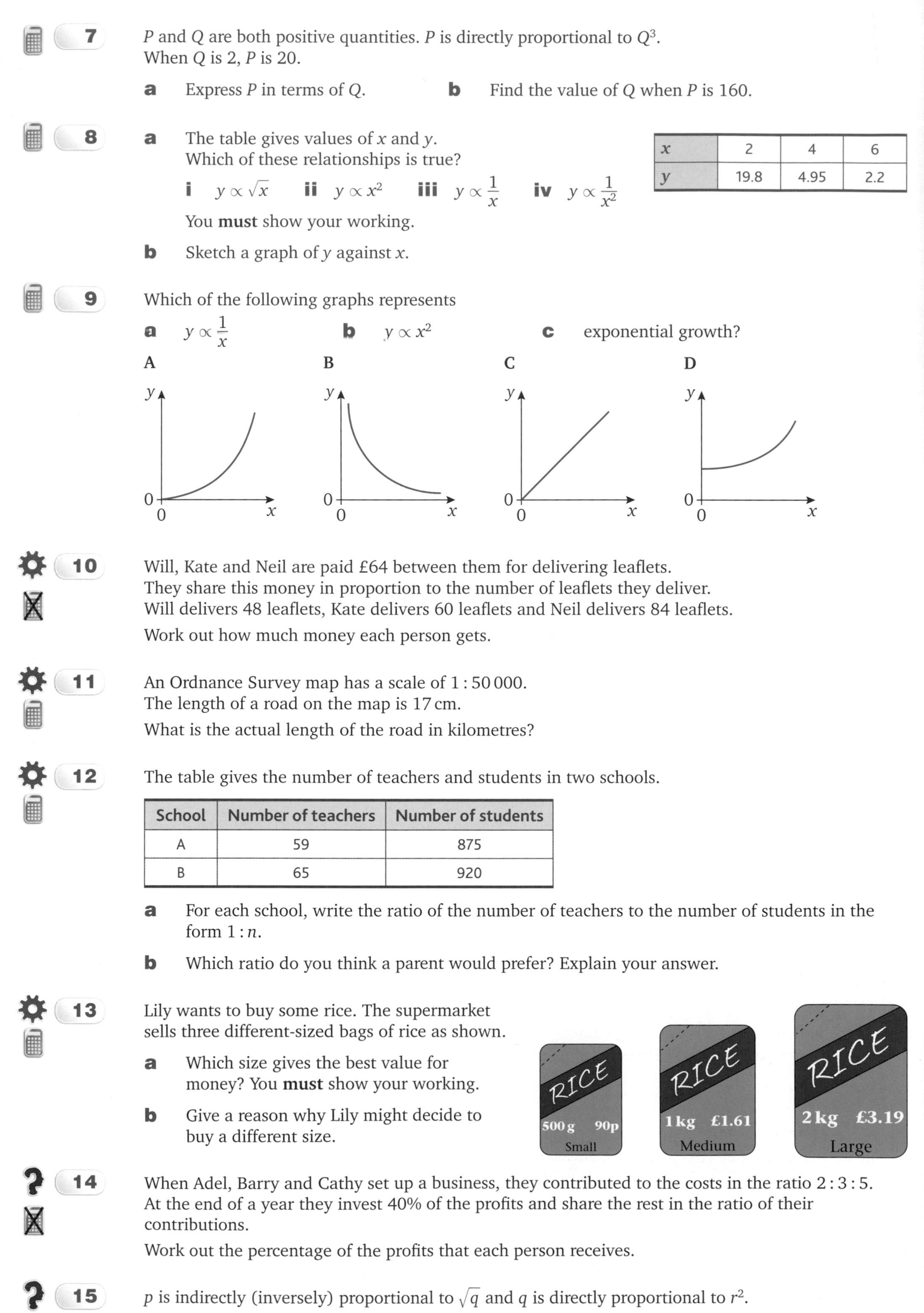

A

7 P and Q are both positive quantities. P is directly proportional to Q^3.
When Q is 2, P is 20.

a Express P in terms of Q. **b** Find the value of Q when P is 160.

8 **a** The table gives values of x and y.
Which of these relationships is true?

i $y \propto \sqrt{x}$ **ii** $y \propto x^2$ **iii** $y \propto \frac{1}{x}$ **iv** $y \propto \frac{1}{x^2}$

You **must** show your working.

x	2	4	6
y	19.8	4.95	2.2

b Sketch a graph of y against x.

A*

9 Which of the following graphs represents

a $y \propto \frac{1}{x}$ **b** $y \propto x^2$ **c** exponential growth?

10 Will, Kate and Neil are paid £64 between them for delivering leaflets.
They share this money in proportion to the number of leaflets they deliver.
Will delivers 48 leaflets, Kate delivers 60 leaflets and Neil delivers 84 leaflets.

Work out how much money each person gets.

11 An Ordnance Survey map has a scale of 1 : 50 000.
The length of a road on the map is 17 cm.

What is the actual length of the road in kilometres?

12 The table gives the number of teachers and students in two schools.

School	Number of teachers	Number of students
A	59	875
B	65	920

a For each school, write the ratio of the number of teachers to the number of students in the form 1 : n.

b Which ratio do you think a parent would prefer? Explain your answer.

13 Lily wants to buy some rice. The supermarket sells three different-sized bags of rice as shown.

a Which size gives the best value for money? You **must** show your working.

b Give a reason why Lily might decide to buy a different size.

14 When Adel, Barry and Cathy set up a business, they contributed to the costs in the ratio 2 : 3 : 5.
At the end of a year they invest 40% of the profits and share the rest in the ratio of their contributions.

Work out the percentage of the profits that each person receives.

15 p is indirectly (inversely) proportional to $\sqrt{q}$ and q is directly proportional to r^2.

Describe the relationship between p and r and sketch a graph of p against r.

Number

5 Indices and standard index form

Key terms

Write down definitions for the following words. Check your answers in the glossary of your Student Book.

cube number
cube root
index
indices
power
square number
square root
standard index form

Revise... Key points

Powers and roots (Units 1 2)

Squares and square roots

The **square numbers** that you should know are:

$1^2 = \mathbf{1}$	$6^2 = \mathbf{36}$	$11^2 = \mathbf{121}$
$2^2 = \mathbf{4}$	$7^2 = \mathbf{49}$	$12^2 = \mathbf{144}$
$3^2 = \mathbf{9}$	$8^2 = \mathbf{64}$	$13^2 = \mathbf{169}$
$4^2 = \mathbf{16}$	$9^2 = \mathbf{81}$	$14^2 = \mathbf{196}$
$5^2 = \mathbf{25}$	$10^2 = \mathbf{100}$	$15^2 = \mathbf{225}$

You should also know the related **square roots**.

If you square a number, you always get a positive result.
For example, $5^2 = 5 \times 5 = 25$ and $(-5)^2 = -5 \times -5 = 25$ also.

AQA *Examiner's tip*

Remember, a negative number multiplied by a negative number gives a positive number.

So when you take the **square root** of 25 you get two answers, 5 and -5.

$\sqrt{25}$ means the positive square root of 25, so $\sqrt{25} = 5$.
The negative square root of 25 is $-\sqrt{25} = -5$

Cubes and cube roots

The **cube numbers** that you should know are:

$1^3 = \mathbf{1}$	$4^3 = \mathbf{64}$
$2^3 = \mathbf{8}$	$5^3 = \mathbf{125}$
$3^3 = \mathbf{27}$	$10^3 = \mathbf{1000}$

Remember that 3^2 means $3 \times 3 = 9$ and 2^3 means $2 \times 2 \times 2 = 8$. Neither of them equals 6!

You should also know the related **cube roots.**

When you cube a negative number, you get a negative result.
For example, $(-5)^3 = -5 \times -5 \times -5 = -125$

When you take the **cube root** of a number, there is just one answer. $\sqrt[3]{125}$ means the cube root of 125.

$\sqrt[3]{125} = 5$ and $\sqrt[3]{-125} = -5$

AQA *Examiner's tip*

Remember that when you take the square root of a positive number there are two possible answers, but when you take the cube root there is just one possible answer.

Higher powers

The **index** or **power** tells you how many times to multiply the base number by itself.

This is the index or power.

So $2^4 = 2 \times 2 \times 2 \times 2 = 16$

This is the base.

and $10^6 = 10 \times 10 \times 10 \times 10 \times 10 \times 10 = 1\,000\,000$ (one million)

In Unit 2 you will have to work out powers without a calculator.
When powers or roots are needed in other units, you can use your calculator.

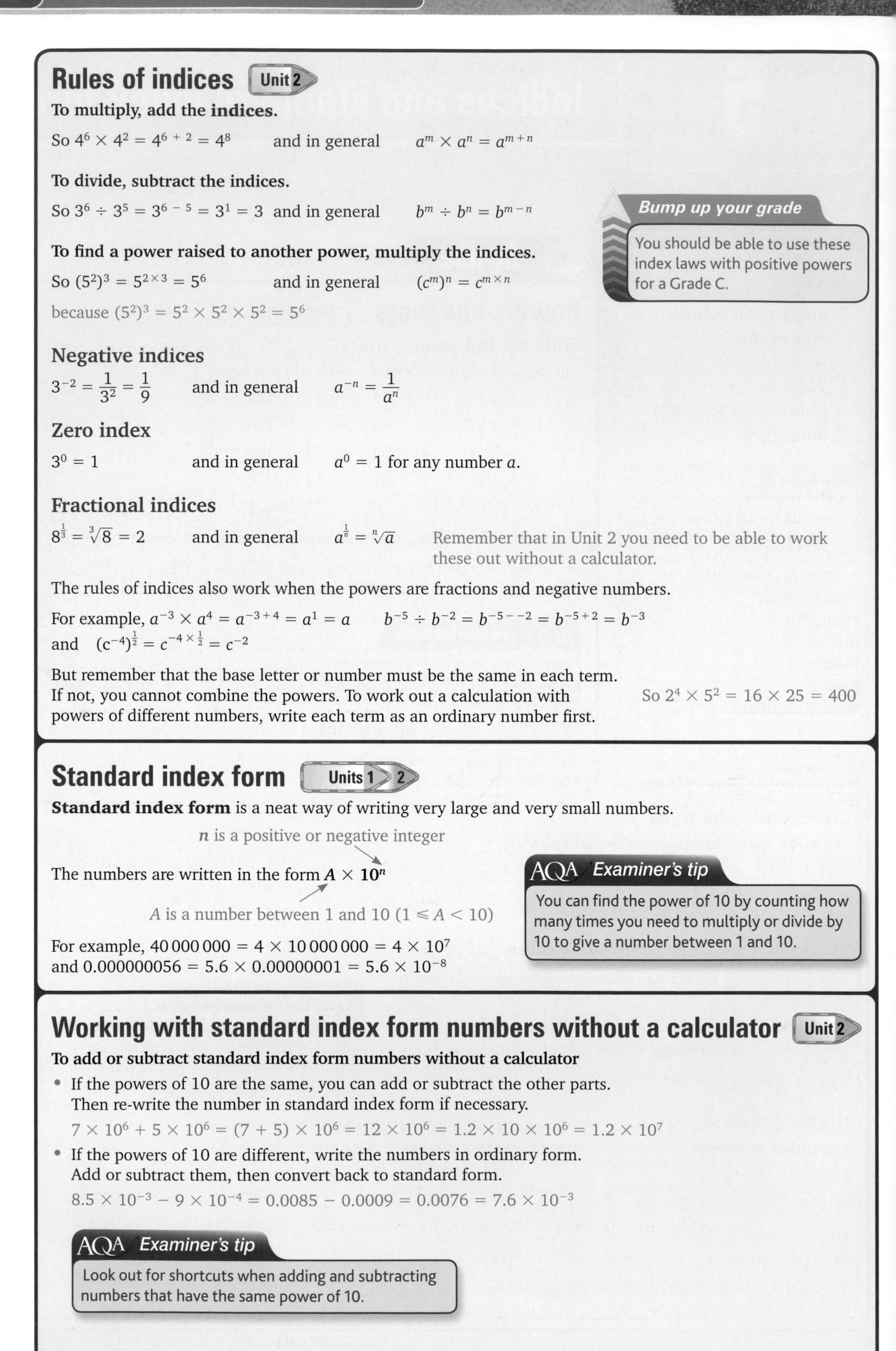

Rules of indices Unit 2

To multiply, add the indices.

So $4^6 \times 4^2 = 4^{6+2} = 4^8$ and in general $a^m \times a^n = a^{m+n}$

To divide, subtract the indices.

So $3^6 \div 3^5 = 3^{6-5} = 3^1 = 3$ and in general $b^m \div b^n = b^{m-n}$

To find a power raised to another power, multiply the indices.

So $(5^2)^3 = 5^{2\times 3} = 5^6$ and in general $(c^m)^n = c^{m\times n}$

because $(5^2)^3 = 5^2 \times 5^2 \times 5^2 = 5^6$

Bump up your grade

You should be able to use these index laws with positive powers for a Grade C.

Negative indices

$3^{-2} = \frac{1}{3^2} = \frac{1}{9}$ and in general $a^{-n} = \frac{1}{a^n}$

Zero index

$3^0 = 1$ and in general $a^0 = 1$ for any number a.

Fractional indices

$8^{\frac{1}{3}} = \sqrt[3]{8} = 2$ and in general $a^{\frac{1}{n}} = \sqrt[n]{a}$ Remember that in Unit 2 you need to be able to work these out without a calculator.

The rules of indices also work when the powers are fractions and negative numbers.

For example, $a^{-3} \times a^4 = a^{-3+4} = a^1 = a$ $b^{-5} \div b^{-2} = b^{-5--2} = b^{-5+2} = b^{-3}$

and $(c^{-4})^{\frac{1}{2}} = c^{-4\times\frac{1}{2}} = c^{-2}$

But remember that the base letter or number must be the same in each term. If not, you cannot combine the powers. To work out a calculation with powers of different numbers, write each term as an ordinary number first.

So $2^4 \times 5^2 = 16 \times 25 = 400$

Standard index form Units 1 2

Standard index form is a neat way of writing very large and very small numbers.

The numbers are written in the form $A \times 10^n$

n is a positive or negative integer

A is a number between 1 and 10 ($1 \leqslant A < 10$)

For example, $40\,000\,000 = 4 \times 10\,000\,000 = 4 \times 10^7$
and $0.000000056 = 5.6 \times 0.00000001 = 5.6 \times 10^{-8}$

AQA Examiner's tip

You can find the power of 10 by counting how many times you need to multiply or divide by 10 to give a number between 1 and 10.

Working with standard index form numbers without a calculator Unit 2

To add or subtract standard index form numbers without a calculator

- If the powers of 10 are the same, you can add or subtract the other parts. Then re-write the number in standard index form if necessary.

  $7 \times 10^6 + 5 \times 10^6 = (7 + 5) \times 10^6 = 12 \times 10^6 = 1.2 \times 10 \times 10^6 = 1.2 \times 10^7$

- If the powers of 10 are different, write the numbers in ordinary form. Add or subtract them, then convert back to standard form.

  $8.5 \times 10^{-3} - 9 \times 10^{-4} = 0.0085 - 0.0009 = 0.0076 = 7.6 \times 10^{-3}$

AQA Examiner's tip

Look out for shortcuts when adding and subtracting numbers that have the same power of 10.

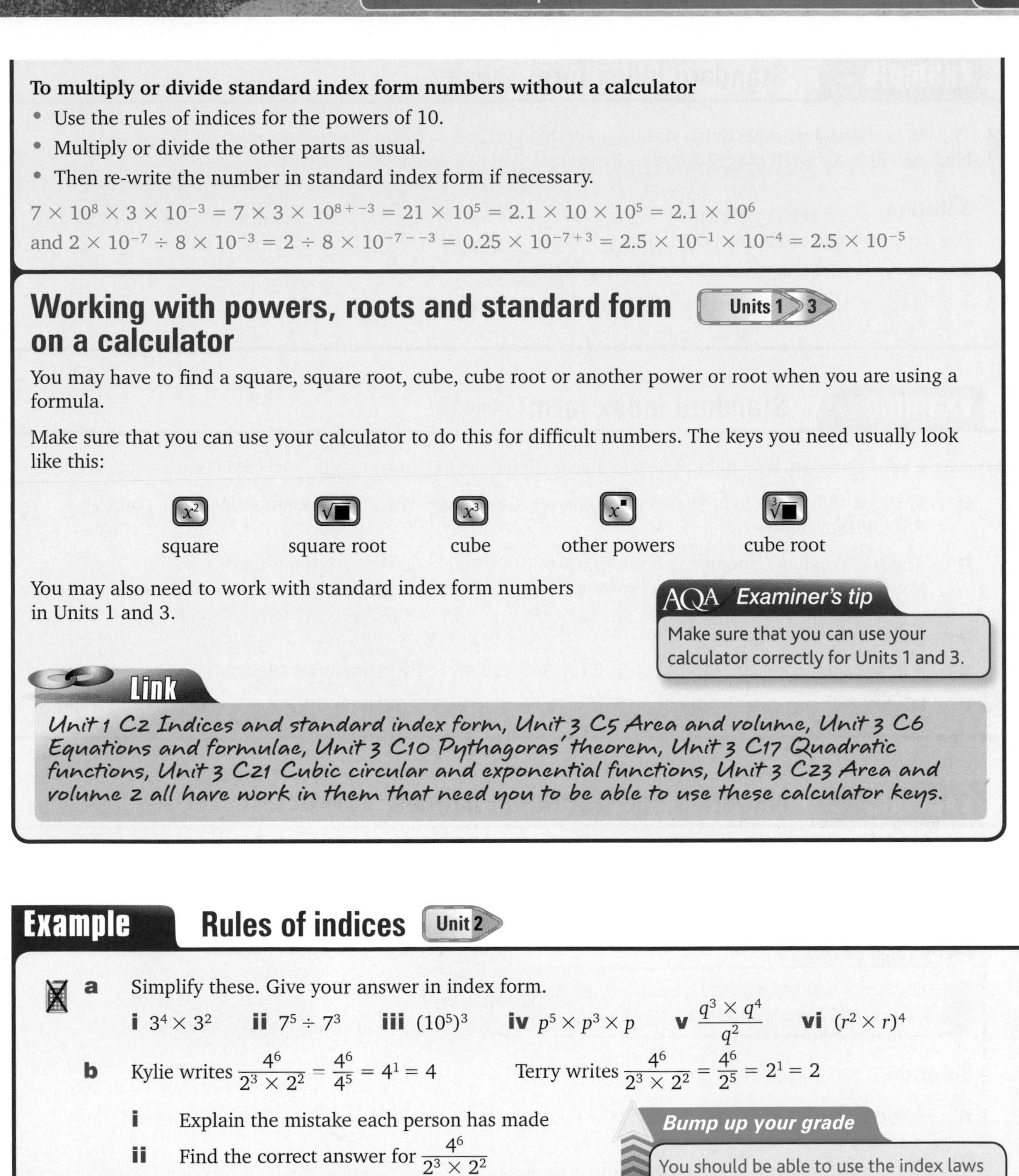

To multiply or divide standard index form numbers without a calculator

- Use the rules of indices for the powers of 10.
- Multiply or divide the other parts as usual.
- Then re-write the number in standard index form if necessary.

$7 \times 10^8 \times 3 \times 10^{-3} = 7 \times 3 \times 10^{8+-3} = 21 \times 10^5 = 2.1 \times 10 \times 10^5 = 2.1 \times 10^6$

and $2 \times 10^{-7} \div 8 \times 10^{-3} = 2 \div 8 \times 10^{-7--3} = 0.25 \times 10^{-7+3} = 2.5 \times 10^{-1} \times 10^{-4} = 2.5 \times 10^{-5}$

Working with powers, roots and standard form on a calculator

Units 1 3

You may have to find a square, square root, cube, cube root or another power or root when you are using a formula.

Make sure that you can use your calculator to do this for difficult numbers. The keys you need usually look like this:

x^2	$\sqrt{\blacksquare}$	x^3	$x^{\blacksquare}$	$\sqrt[3]{\blacksquare}$
square	square root	cube	other powers	cube root

You may also need to work with standard index form numbers in Units 1 and 3.

AQA Examiner's tip

Make sure that you can use your calculator correctly for Units 1 and 3.

Link

Unit 1 C2 Indices and standard index form, Unit 3 C5 Area and volume, Unit 3 C6 Equations and formulae, Unit 3 C10 Pythagoras' theorem, Unit 3 C17 Quadratic functions, Unit 3 C21 Cubic circular and exponential functions, Unit 3 C23 Area and volume 2 all have work in them that need you to be able to use these calculator keys.

Example Rules of indices

Unit 2

C

a Simplify these. Give your answer in index form.

i $3^4 \times 3^2$ **ii** $7^5 \div 7^3$ **iii** $(10^5)^3$ **iv** $p^5 \times p^3 \times p$ **v** $\frac{q^3 \times q^4}{q^2}$ **vi** $(r^2 \times r)^4$

b Kylie writes $\frac{4^6}{2^3 \times 2^2} = \frac{4^6}{4^5} = 4^1 = 4$ Terry writes $\frac{4^6}{2^3 \times 2^2} = \frac{4^6}{2^5} = 2^1 = 2$

i Explain the mistake each person has made

ii Find the correct answer for $\frac{4^6}{2^3 \times 2^2}$

Bump up your grade

You should be able to use the index laws with positive powers for a Grade C.

Solution

a **i** $3^4 \times 3^2 = 3^{4+2} = 3^6$

ii $7^5 \div 7^3 = 7^{5-3} = 7^2$

iii $(10^5)^3 = 10^{5 \times 3} = 10^{15}$

iv $p^5 \times p^3 \times p = p^{5+3+1} = p^9$ p is the same as p^1

v $\frac{q^3 \times q^4}{q^2} = q^{3+4-2} = q^5$

vi $(r^2 \times r)^4 = (r^3)^4 = r^{12}$

b **i** Kylie has multiplied the base numbers in the denominator. $2^3 \times 2^2 = 2^5$ **not** 4^5

Terry is correct up to $\frac{4^6}{2^5}$, but then he has divided the base numbers.

ii As the base numbers are different, you cannot use the rules of indices.

$$\frac{4^6}{2^5} = \frac{4 \times 4 \times 4 \times 4 \times 4 \times 4}{2 \times 2 \times 2 \times 2 \times 2} = \frac{\cancel{2} \times \cancel{2} \times \cancel{2} \times \cancel{2} \times \cancel{2} \times 2 \times 2 \times 2 \times 2 \times 2 \times 2 \times 2}{\cancel{2} \times \cancel{2} \times \cancel{2} \times \cancel{2} \times \cancel{2}} = 2 \times 2 \times 2 \times 2 \times 2 \times 2 \times 2 = 2^7 \text{ or } 128$$

Example: Standard index form (Unit 2)

B

The UK population is expected to rise from approximately 6×10^7 to 7×10^7 in the next 20 years. Find the expected percentage increase. Give your answer to the nearest per cent.

Solution

The expected increase in the population $= 7 \times 10^7 - 6 \times 10^7 = (7 - 6) \times 10^7 = 1 \times 10^7$

The expected percentage increase in the population

$= \frac{1 \times 10^7}{6 \times 10^7} \times 100\% = \frac{1}{6} \times 100\% = 17\%$ to the nearest per cent.

$10^7 \div 10^7 = 10^0 = 1$

Example: Standard index form (Unit 1)

B

A light year is the distance light travels in a year (365 days). Light travels at 3×10^5 kilometres per second.

a Work out how far a light year is in kilometres. Give your answer in standard index form, correct to 3 significant figures (s.f.).

b The star Proxima Centauri is about 40 200 000 000 000 kilometres from the Earth. Work out how many light years Proxima Centauri is from Earth. Give your answer to the nearest integer.

Solution

a 1 light year $= 3 \times 10^5 \times 60 \times 60 \times 24 \times 365 = 9.46 \times 10^{12}$ kilometres (to 3 s.f.).

b Distance in light years $= (4.02 \times 10^{13}) \div (9.46 \times 10^{12}) = 4.249... = 4$ light years (to the nearest integer).

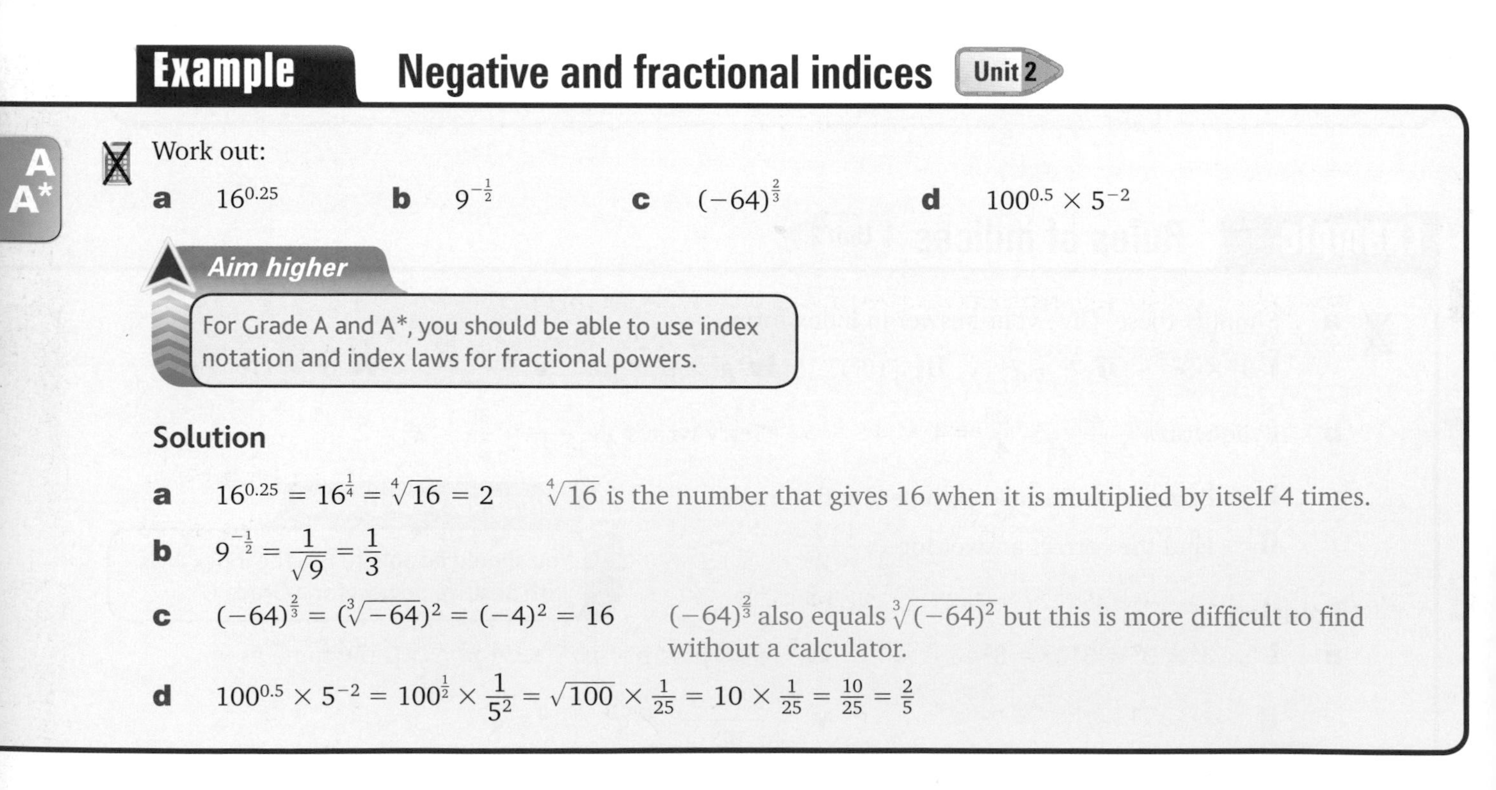

Example: Negative and fractional indices (Unit 2)

A A*

Work out:

a $16^{0.25}$ **b** $9^{-\frac{1}{2}}$ **c** $(-64)^{\frac{2}{3}}$ **d** $100^{0.5} \times 5^{-2}$

Aim higher

For Grade A and A*, you should be able to use index notation and index laws for fractional powers.

Solution

a $16^{0.25} = 16^{\frac{1}{4}} = \sqrt[4]{16} = 2$ $\sqrt[4]{16}$ is the number that gives 16 when it is multiplied by itself 4 times.

b $9^{-\frac{1}{2}} = \frac{1}{\sqrt{9}} = \frac{1}{3}$

c $(-64)^{\frac{2}{3}} = (\sqrt[3]{-64})^2 = (-4)^2 = 16$ $(-64)^{\frac{2}{3}}$ also equals $\sqrt[3]{(-64)^2}$ but this is more difficult to find without a calculator.

d $100^{0.5} \times 5^{-2} = 100^{\frac{1}{2}} \times \frac{1}{5^2} = \sqrt{100} \times \frac{1}{25} = 10 \times \frac{1}{25} = \frac{10}{25} = \frac{2}{5}$

Practise... Indices and standard index form

D C B A A*

D

1 Which of each pair is the greater?

a 0.9^2 or 0.9^3 **b** 0.36 or $\sqrt{0.36}$ **c** $-\sqrt{25}$ or $\sqrt[3]{-27}$

D

2 Write down approximate values for each of these.

a 6.9^2 **d** $(-0.995)^3$
b $(-3.2)^2$ **e** the square root of 83
c 4.93^3 **f** $\sqrt[3]{-7.95}$

Bump up your grade

For a Grade C, you should be able to find the cube root of a negative number. You should also know that a positive number has a positive and a negative square root.

3 Work out the value of each of the following.

a 10^4 **b** $(-10)^4$ **c** 2^5 **d** $(-2)^5$ **e** 5^4 **f** 4^5

C

4 Use the rules of indices to simplify the following. Give your answers in index form.

a $3^5 \times 3^4$ **d** $p^7 \times p^3$ **g** $\frac{5^8}{5^6}$ **j** $\frac{9^3 \times 9^2}{9}$
b $7^6 \div 7^2$ **e** $q^4 \div q^3$ **h** $(4^5 \times 4^2)^3$ **k** $\frac{t^8}{t^2 \times t^5}$
c $(6^3)^2$ **f** $(r^3)^5$ **i** $(8^7 \div 8)^4$ **l** $\left(\frac{n^6}{n^4}\right)^3$

5 Work these out. Give your answers as ordinary numbers.

a $2^4 \times 3^2$ **b** $10^3 \div 5^2$ **c** $\frac{9^3}{3^4}$ **d** $\frac{2^3 \times 5^2}{10^2}$ **e** $\frac{10^4}{2^3}$

6 Chloe says that $3^4 \times 3^3 = 3^{12}$. Dave says $3^4 \times 3^3 = 9^7$

a Explain the error that each person has made.
b What is the correct answer?

B

7 **a** Write these numbers in standard index form.

i 8 560 000 **ii** 0.000075

b Write these as ordinary numbers.

i 2.6×10^8 **ii** 6.24×10^{-3}

8 Write the following numbers in order of size. Start with the smallest.

2.4×10^6 1.95×10^7 6.8×10^{-2} 9.3×10^{-4} 5.7×10^{-2} 1.82×10^{-3}

9 Work these out. Give your answers in standard index form.

a $(4 \times 10^5) + (2 \times 10^5)$ **c** $(6 \times 10^4) + (8 \times 10^4)$ **e** $(3 \times 10^3) \div (5 \times 10^4)$
b $(7 \times 10^{-3}) - (5 \times 10^{-3})$ **d** $(3 \times 10^3) \times (5 \times 10^4)$ **f** $(3.6 \times 10^{13}) \div (2 \times 10^{-7})$

B A

10 Use the rules of indices to simplify the following.

a $7^{-7} \times 7^6$ **d** $p^{-2} \times p^{-3}$ **g** $\frac{10^4 \times 10^5}{10^3}$ **j** $(3y)^0$
b $2^5 \div 2^{-3}$ **e** $b^{-4} \div b^2$ **h** $(y \times y^{-4})^3$ **k** $4r^{-1} \times 2r^{-2}$
c $(5^{-1})^3$ **f** $(n^{-3})^{-2}$ **i** $5a^2 \times 2a^3$ **l** $\frac{6p^3 \times 3p}{9p^2}$

Hint

In part **i** write $5a^2 \times 2a^3$ as $5 \times 2 \times a^2 \times a^3 = 10 \times a^2 \times a^3$

A A*

11 Work out:

a $27^{\frac{1}{3}}$ **b** $5x^0$ **c** $1000^{\frac{2}{3}}$ **d** $8^{-\frac{2}{3}}$ **e** $(3ab)^0$ **f** $16^{-\frac{3}{4}}$

12 Julie is buying a new carpet.
The diagram shows its dimensions.
The carpet costs £10.99 per square metre and the underlay costs £4.99 per square metre. She pays £4 per square metre for having the carpet fitted.

Estimate the total cost.

4 m
4 m
1.9 m
1.9 m

13 Callisto, Europa, Ganymede and Io are moons of the planet Jupiter. The table gives the distance between each of these moons and Jupiter.

Moon	Distance from Jupiter (km)
Callisto	1.88×10^6
Europa	6.71×10^5
Ganymede	1.07×10^6
Io	4.22×10^5

a Write the moons in order of how far they are from Jupiter. Start with the nearest.

b Assume that Io travels in a circle of radius 4.22×10^5 kilometres around Jupiter.
Work out how far Io travels in one revolution.
Give your answer
i as an ordinary number to the nearest 10 000 km.
ii in standard index form to 2 significant figures.

14 **a** Write these in standard index form: **i** £880 million **ii** 4.4 billion

b A newspaper says that people have downloaded 4.4 billion mobile apps this year at a total cost of £880 million. Work out the average cost per download. Give your answer in pence.

15 The population of England last year was approximately 50 million.

a Write this number in standard index form.

b The number of Key Stage 4 pupils was approximately 6×10^5.
What percentage is this of England's population?

16 The table gives some of a magazine's predictions for the money that will be spent on downloading mobile applications in 2013.

Type	Predicted revenue
Music	3.2×10^9 euro
Games	2.8×10^9 euro
Video	1.8×10^9 euro

a Work out the total predicted revenue.

b Calculate the percentage from each type of application.
Give your answers to the nearest per cent.

17 When you square any positive integer that has an 'end digit' of 1, the answer has the same 'end digit' (that is, the 'end digit' of the square is also 1).

For example, $11^2 = 121$, $21^2 = 441$, $31^2 = 961$ and so on.

a Explain why this occurs.

b Find three other 'end digits' for which the same is true.

18 **a** Work out: **i** $1 + 3$ **ii** $1 + 3 + 5$ **iii** $1 + 3 + 5 + 7$

b Find the sum of the first 100 odd numbers.

19 A chessboard has 64 squares. Half of them are black and the other half are white.

a The sides of the squares on one chessboard are all 2 cm long.
Write each of the following as a power of 2:
i the length of the sides of the chessboard
ii the area of one square
iii the area of the chessboard
iv the total area of the black squares.

b According to a legend, a wise man asked his emperor for the following payment for inventing the game of chess:
1 grain of rice on the first square of the chessboard,
2 grains of rice on the second square, 4 grains of rice on the third square and so on (with the number of grains doubled from each square to the next).

According to this rule, how many grains of rice would there be on the last square of the chessboard?

AQA Examination-style questions

1 a Rachel earns £18 400 per year. She gets a pay rise of 3%.
How much is her pay now? *(2 marks)*

b Steve earns £16 250 per year. He gets a pay rise of £520.
Write this as a percentage. *(2 marks)*

2 £2400 is shared in the ratio 7 : 3 : 2. Work out the largest share. *(3 marks)*

3 Three shops advertise the same bookcase.

Shop P	Shop Q	Shop R
£40 deposit Plus 6 equal monthly payments of £15	Usual price £165 Special offer $\frac{1}{3}$ off usual price	Usual price £140 Special offer 20% off usual price

At which shop is the bookcase cheapest? You **must** show your working. *(5 marks)*

4 Steve says, 'When you add two prime numbers together you always get an even number.'
Give an example to show that Steve is not correct. *(2 marks)*

5 Write down all the common factors of 28 and 42. *(2 marks)*

6 a Work out: **i** 0.8×0.2 *(1 mark)*
ii $2.4 \div 0.3$ *(1 mark)*

b Estimate the value of: **i** 31.9×591 *(2 marks)*
ii $159.3 \div 21.95$ *(2 marks)*

7 a Find both square roots of 4624. *(2 marks)*

b Calculate the cube of 1.7. *(1 mark)*

c Calculate $\frac{86.9 - 12.7}{2.4^2}$

i write down your full calculator display *(1 mark)*

ii write your answer to 1 decimal place. *(1 mark)*

d Write down the reciprocal of 64. *(1 mark)*

D

D C

C

8 Rosie has two dogs, Bill and Ben.
Bill eats $\frac{1}{2}$ of a tin of dog food each day. Ben eats $\frac{2}{3}$ of a tin of dog food each day.
What is the smallest number of tins of dog food that Rosie will need to feed both dogs for one week? *(4 marks)*

9 **a** Work out:
i $4\frac{8}{9} - 1\frac{5}{6}$ *(3 marks)*
ii $4\frac{8}{9} \div 1\frac{5}{6}$ *(3 marks)*
b Write down the reciprocal of 0.25 *(1 mark)*

10 **a** Write 72 as a product of its prime factors. Give your answer in index form. *(2 marks)*
b Write these numbers as the product of their prime factors in index form.
i 144 *(1 mark)*
ii 720 *(1 mark)*

11 There are three bus routes across a town.
The number 52 bus leaves the station every 12 minutes, the number 42 bus leaves the station every 15 minutes and the number 86 bus leaves the station every 20 minutes.
The buses have just left the station together. How many times will they leave the station together in the next 12 hours? *(3 marks)*

12 The normal price of a 250 gram pack of butter is 90p. There are two special offers.

Offer A 30% off the normal price	**Offer B** 30% extra butter. Price still 90p.

Which offer is best value for money?
You **must** show your working. *(5 marks)*

13 In a sale the price of a computer is reduced from £600 to £480.
What is the percentage reduction? *(3 marks)*

14 Sally wants to make some coconut cookies.
She only has 100 g of butter.
How much coconut should she use? *(3 marks)*

Coconut Cookies
125 g butter
100 g of sugar
100 g of flour
75 g of oats
75 g of coconut

C B

15 **a** Simplify: **i** $t^3 \times t^9$ **ii** $t^3 \div t^9$ **iii** $(t^3)^9$ *(3 marks)*
b If $t = -1$, which answer in part **a** is negative? *(1 mark)*
c If $t = 0.1$, which answer in part **a** has the greatest value? *(1 mark)*

16 **a** Which of the following fractions is nearest to $\frac{2}{3}$?
$\frac{3}{5}$ $\frac{3}{4}$ $\frac{13}{20}$
You **must** show your working. *(2 marks)*
b Write $0.0\dot{2}\dot{4}$ as a fraction. Give your answer in its simplest form. *(2 marks)*

B

17 Money left in a building society account earns interest at a fixed rate of 5.4% per year. Crystal invests £6000 in this account.

a Work out how much is in the account after 5 years. *(3 marks)*

b Show that it will take ten years for the amount to grow to more than £10 000. *(3 marks)*

18 a Explain why 27.5×10^{-3} is **not** written in standard index form. *(1 mark)*

b Work out $\frac{9 \times 10^9}{3 \times 10^3}$. Give your answer in standard index form. *(2 marks)*

c Find the difference between 6.4×10^{-2} and 1.2×10^{-3}
Give your answer as an ordinary number. *(2 marks)*

A

19 The mass of Neptune is 1.02×10^{26} kg and the mass of the Earth is 5.97×10^{24} kg.

a How many times heavier is Neptune than the Earth?
Give your answer to the nearest integer. *(2 marks)*

b Neptune is approximately a sphere of radius 2.47×10^7 metres.

i Use the formula $V = \frac{4}{3}\pi r^3$ to calculate the approximate volume of Neptune.
Give your answer in standard index form in cubic metres to 3 significant figures. *(3 marks)*

ii Find the approximate density of Neptune in kg/m^3 to 2 significant figures. *(2 marks)*

20 Two variables, x and y, are connected by the relationship 'y is indirectly proportional to the square of x'.

a When $x = 5$, $y = 20$. Express y in terms of x. *(3 marks)*

b Explain what happens to the value of y when the value of x is doubled. *(2 marks)*

c Sketch a graph of y against x. *(2 marks)*

A
A*

21 a Simplify fully $\frac{(\sqrt{5})^3}{10}$ *(2 marks)*

b Write $\frac{6}{\sqrt{3}} + \sqrt{27}$ in the form $k\sqrt{3}$ *(4 marks)*

22 a Work out $3^4 \times 81^{-0.5}$ *(3 marks)*

b If $27^x = \frac{1}{9}$ find the value of x. *(2 marks)*

23 A report says that in Europe there are 33 million mobile broadband connections. It predicts that this number will increase by 46% per year. Assume this occurs.

a How many mobile broadband connections will there be 4 years from now? *(3 marks)*

b i Sketch a graph of the number of mobile broadband connections against time. *(2 marks)*

ii Describe the growth in mobile broadband connections. *(1 mark)*

iii Give a reason why this growth is unlikely to continue for the next 50 years. *(1 mark)*

1 Collecting data

Unit 1

3 Collecting data

Key terms

Write down definitions for the following words. Check your answers in the glossary of your Student Book.

census
closed questions
continuous data
controlled experiment
data collection sheet
data logging
discrete data
hypothesis
observation
observation sheet
open questions
pilot survey
population
primary data
qualitative data
quantitative data
questionnaire
random sampling
raw data
sample
sample size
secondary data
stratified (random) sampling
survey
two-way table

Revise… Key points

Data handling cycle Unit 1

The data handling cycle is the framework for work in statistics. It has four stages.

The main reason for collecting data is to investigate a **hypothesis**. A hypothesis is a statement that you want to investigate (for example Do Year 10 girls take longer to get ready in the morning than Year10 boys?). Specifying a hypothesis is the first stage of the data handling cycle.

Link

Chapter 3 of the Student Book, Learn 3.1, clearly illustrates the whole data handling cycle with a flow diagram.

Collecting the data you need is the second stage of the cycle. The data that you first collect are called **raw data**. Raw data are data before they have been sorted.

Types of data Unit 1

Data can be primary or secondary

Primary data are data that you collect yourself in order to investigate a hypothesis. For example, they may be data that you collected in a science experiment or a questionnaire you used on students.

Secondary data are data that have already been collected by someone else. For example, these may be data you have found on the internet or in a reference book.

Data can be qualitative or quantitative

Qualitative data are non-numerical data, for example hair colour or method of transport to get to work.

Quantitative data are data that take numerical values, for example shoe size or height.

Quantitative data can be discrete or continuous

Discrete data are data that can only take a set of fixed values, for example shoe size or the number of days absent from work last year.

Continuous data are data that can take any value within a range, for example the height of a Year 7 student or the speed of a free kick taken by a footballer.

Data collection methods Unit 1

A **questionnaire** is a common method for collecting data. A questionnaire consists of a number of questions. Questions may be **open** or **closed**.

Open questions allow for any response (answer) to be given (for example 'What did you have for breakfast?').

Closed questions have a set of responses that can be chosen. (See below for an example.)

All the possible responses to a question are put into a response section.

How many singles have you downloaded in the last week?

0 ☐   1–2 ☐   3–5 ☐   6–10 ☐   11+ ☐

This is an example of a closed question.

This is the response section.

You may be asked in an exam to comment on a question or response section from a questionnaire.

The table below shows some question and response sections and identifies the possible problems.

Question and response section	Problem
How old are you?	This question is too personal. If you want to ask this question you need to create a response section like this: 16 to 25, 26 to 35, 36 to 50, etc.
Don't you agree that fast food is unhealthy for you? Yes ☐ No ☐ Don't know ☐	This question is biased and is trying to encourage you to answer in a particular way.
How often do you exercise? Rarely ☐ Now and again ☐ Often ☐ Nearly every day ☐	The problem with this question is that there is no period of time indicated (e.g. in a week, month, etc.). The response section is also very vague and could be interpreted by people in different ways.
How many hours of TV did you watch last night? 0–1 ☐ 1–3 ☐ 3–5 ☐	Some of the response options are overlapping (e.g. where would you put 1 hour?). Also there is no option for over 5 hours.
Do you think more should be done to tackle global warming? Yes ☐ No ☐	This question requires a 'don't know' or 'no opinion' option box. This is for people who do not have an opinion on the matter or do not have enough information to be able to answer the question.

Raw data generated from a questionnaire are often organised in the form of a table.

This could be a frequency table or **two-way table**.

Link

Chapter 3 of the Student Book, Learn 3.2, contains a useful table that summarises other methods of collecting data.

Bump up your grade

To get a Grade C, you need to be confident working with questionnaires. Learn the different errors and problems that can appear on questionnaires.

Sampling methods

Simple **random sampling** is where every member of a population has the same chance of being chosen. If, for example, you were asked to choose 30 students from Year 11 to take part in a survey, you could put every student's name into a hat and take out 30 names. Alternatively, you could assign each student in Year 11 a number and then use random number tables or the random number button on your calculator.

Stratified random sampling can be used when the population is split into different 'categories' (called strata), for example males and females. In stratified sampling, a set number of people or items are chosen from each category. The number of people or items chosen from each category is in the same proportion as they are in the whole population. Stratified sampling is used to ensure that the sample obtained is as representative of the population as possible.

Example Questionnaires Unit 1

C

Abdul and his family are travelling by train to London.

The customer service manager asks passengers to complete a questionnaire.

Here is one of the questions.

How often do you travel with us and use our restaurant?

Rarely ☐ Occasionally ☐ Often ☐ Very often ☐

a Write down one criticism of the question asked.

b Write down one criticism of the response section.

c How could the question and response section be improved?

Solution

a The question is made up of two questions.
It is possible that you travel with the company, but do not use the restaurant.

b The response section is very vague.
Different people may interpret the responses in different ways.

c The company could ask two separate questions, for example:

i In the past month how many times have you travelled with us?

1–2 ☐ 3–5 ☐ 6–10 ☐ 11+ ☐

ii Have you used our restaurant services? Yes ☐ No ☐

Example Stratified sampling Unit 1

A

1 The table below shows the ages of members of a chain of local health and fitness clubs.

Age	10–18	19–25	26–35	36–50	51+
Number of members	259	785	1850	955	151

The owner of the clubs wishes to ask the members about the services on offer.

One method suggested is to ask all the members of the club.

a Comment on this method of survey.

An alternative method suggested is to ask 50 members from each of the age groups.

b Comment on this method.

The owner decides to ask 200 members in total using stratified sampling.

c How many members should be chosen from each age group?

2 A mail order company wishes to survey its customers about new products they would like to see sold. The company knows from its records that 22% of its customers are male.

The company sends its questionnaire to a sample of 500 male customers.

How many female customers should it send the questionnaire to?

Solution

1 a Asking all the members will get lots of views.

However, the method is likely to be too time-consuming and expensive.

It will also take a long time to analyse the data.

b Asking 50 members from each of the age groups will ensure that you get views from members in each category.

However, some age groups would be over-represented.

You would be asking a greater proportion of the 51+ age group than the 26–35 age group.

This could lead to results that are not representative of the overall population.

c First find the total number of members.

$259 + 785 + 1850 + 955 + 151 = 4000$

Now divide the sample size required (200) by the total sample (4000) and then multiply by 100.

$$\frac{200}{4000} \times 100 = 5\%$$

This means we are going to take a 5% sample from each age group.

10–18	$0.05 \times 259 = 12.95$	(Rounded to 13 members)
19–25	$0.05 \times 785 = 39.25$	(Rounded to 39 members)
26–35	$0.05 \times 1850 = 92.5$	(Rounded to 93 members)
36–50	$0.05 \times 955 = 47.75$	(Rounded to 48 members)
51+	$0.05 \times 151 = 7.55$	(Rounded to 8 members)

AQA Examiner's tip

Always check that your rounded answers add up to the sample size required.

Now check that the total sample will be 200.

$13 + 39 + 93 + 48 + 8 = 201$

The sample chosen is one person too big. This is due to rounding the numbers.

You will need to reduce one of the samples.

You should reduce the 26–35 age category to 92 people as this is the decimal that was closest to being rounded down.

The number of members to be chosen from each age group is

10–18 (13 members) 19–25 (39 members) 26–35 (92 members)
36–50 (48 members) 51+ (8 members)

2 Assuming the company use stratified sampling:

Males represent 22% of the mail order company's customers.

You need to work out how many questionnaires 1% represents.

500 questionnaires must be 22% of the total number of questionnaires that were sent out, therefore 1% must be

$$\frac{500}{22} = 22.727272\ldots$$

Females represent 100% – 22% = 78% of the customers.

The number of questionnaires sent to females must equal

$$\frac{500}{22} \times 78 = 1772.7272\ldots$$

Therefore 1773 questionnaires must have been sent to females.

Aim higher

To get a Grade A, you need to understand stratified sampling. If you are given information about the sample size of one category, you need to know how to find the sample sizes for the other categories.

Practise... Collecting data Unit 1

D C A

D

1 A local council wants to find out about people's opinions on their services. They want to ask 200 people. Three possible ideas for conducting a survey are discussed.

Method 1	Method 2	Method 3
Visit the local shopping centre on Monday morning and ask 200 people.	Get the local phone book and ring 200 people at random.	Carry out a door-to-door survey in one particular area of town.

a Write down a criticism for each method of collecting data.

b Suggest a better way that the town council might get people's opinion on their services.

C

2 In a town there are two supermarkets, Pricewise and Freshfoods.

Pricewise supermarket is carrying out a survey.

A member of staff stands at the entrance to their store and asks shoppers at random. The shoppers are asked the following question:

Do you agree that our store provides better value for money than Freshfoods?

Strongly agree ☐ Agree ☐ Don't know ☐

a Write down one criticism of the question.

b Write down one criticism of the response section.

c Write down one criticism of the method of selection.

d Is the information collected an example of primary or secondary data? Explain your answer.

3 You have been asked by your local council to help them determine what facilities a new leisure centre should contain.

They would like you to advise them on how to find out people's opinions.

Write a short report to the council to advise them on how to determine the facilities that people in the town would like to see.

You should make reference in your report to the full data handling cycle.

Try and include comments:

- on how the council should collect the data
- on how they should organise and analyse the data once collected
- on how they should make their conclusions.

A

4 The number of students studying various courses at a particular college is shown below.

Course	Science	English	Mathematics	Psychology
Number of students	405	294	186	115

The librarian wishes to carry out a survey about library usage.

She wishes to ask 80 students in total.

She is going to use stratified sampling.

How many students should she pick from each course?

A

5 Two classes study GCSE PE.

The two-way table below shows how many boys and girls are in each class.

	Boys	Girls
Class A	11	18
Class B	24	7

The head of PE wants to carry out a stratified sample.

He needs to choose 15 students in total.

a Explain the benefits of a stratified sample.

b What percentage of class A are boys?

c Determine how many boys and girls from each class should be chosen.

d Describe briefly how the students can then be selected.

6 A train company wishes to survey its drivers and conductors regarding new working arrangements. The company employs 150 drivers and 840 conductors.

The company chooses to use the method of stratified sampling.

In the sample they survey 20 drivers.

How many conductors should they pick?

7 A factory that makes breakfast cereals has two machines, A and B.

Each machine is designed to pack cereal into boxes.

The company produces three different types of cereal.

The table below shows how many boxes of each type of cereal each machine packs in one hour.

	Cereal 1	Cereal 2	Cereal 3
Machine A	950	1780	5900
Machine B	180	4570	3250

Each hour, a sample of the boxes is checked to ensure that the contents measure the correct weight.

The company wishes to select 120 boxes of cereal in total.

How many boxes of Cereal 2 from Machine A should be checked?

2 Statistical measures

Unit 1
6 Statistical measures

Key terms

Write down definitions for the following words. Check your answers in the glossary of your Student Book.

- average
- class interval
- continuous data
- discrete data
- frequency distribution
- grouped data
- inter-quartile range
- mean
- median
- modal class or modal group
- mode
- range

Revise... Key points

Discrete frequency distributions (Unit 1)

A frequency distribution shows how often individual values occur.

A frequency distribution is usually presented in the form of a **frequency table**.

Frequency distributions are usually used with **discrete data**.

Measures of average

When the data are presented in the form of a frequency table, the **mean** is given by the formula

$$\text{mean} = \frac{\text{the total of (frequencies} \times \text{values)}}{\text{the total of frequencies}} = \frac{\Sigma fx}{\Sigma f}$$

where Σ means the sum of

Usually the x values are in the first column (or row) and the frequencies, f, are in the second column.

The **mode** is the value that appears the most. In other words, it is the value that has the highest frequency.

The **median** is the middle value when the values are listed in order.

The median value is the $\frac{1}{2}(n + 1)$th value, where n is the total number of values (that is, the total frequency).

Measures of spread

The **range** is the difference between the highest value and the lowest value.

The **inter-quartile range** is found using

inter-quartile range = upper quartile − lower quartile

The lower quartile is the $\frac{1}{4}(n + 1)$th value.

The upper quartile is the $\frac{3}{4}(n + 1)$th value.

The range and inter-quartile range should be given as single values and not as a range of values.

The inter-quartile range is sometimes preferred to the range as it eliminates any unusually large or small values (called outliers).

Grouped frequency distributions (Unit 1)

Grouped frequency distributions are usually used with **continuous data**.

Measures of average

When the values in the first column (or row) are grouped into **class intervals**, an estimate of the mean is calculated using the formula

$$\text{mean} = \frac{\text{the total of (frequencies} \times \text{midpoint)}}{\text{the total of frequencies}} = \frac{\Sigma fx}{\Sigma f}$$

This is only an estimate of the mean, as we do not know the original exact values.

The **modal class** is the class or group that corresponds to the highest frequency.

The class interval that contains the **median** is the interval that contains the $\frac{1}{2}(n + 1)$th value.

Measures of spread

The **range** is estimated by finding the difference between the highest and lowest values in the table. Because you do not know the exact highest and lowest values, you would subtract the lowest value in the first class interval from the highest value in the last class interval.

For an approximate calculation of the median and to find the **inter-quartile range** for a grouped distribution, you would usually use a cumulative frequency diagram.

See Chapter 3 Representing data for an example of a cumulative frequency diagram.

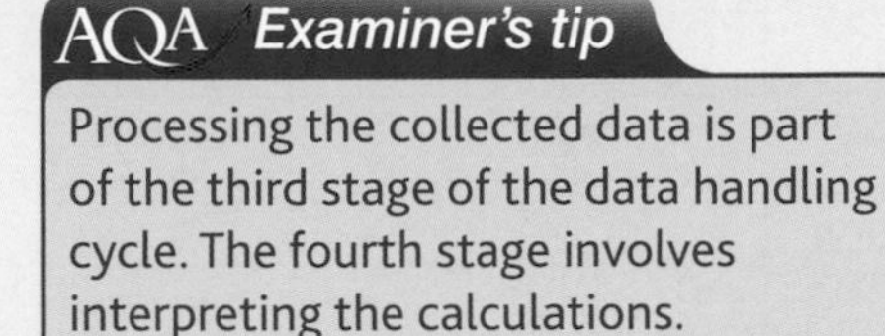

AQA Examiner's tip

Processing the collected data is part of the third stage of the data handling cycle. The fourth stage involves interpreting the calculations.

Example Frequency distributions Unit 1

D

1 Mr George teaches class 9R one lesson per week.

There are 32 students in the class.

Mr George records the number of students absent from his lesson each week over the whole school year.

The results are shown in the table.

Number of students absent	Number of weeks
0	5
1	14
2	12
3	4
4	2
5	1
6	1

a Write down the modal number of students absent.

b Find the median number of students absent.

c Find the mean number of students absent.

d Use your answer to part **c** to calculate the mean number of students present.

e Find the range of students absent.

Solution

1 Begin by thinking of the first column as the x values, and the second column as the list of frequencies, f.

a The mode is the value that corresponds to the highest frequency.

The highest frequency is 14, which corresponds to the value of 1 student.

Therefore, the modal number of students absent is 1 student.

AQA Examiner's tip

Be careful that you do not write down the frequency that corresponds to the mode as your answer.

b The median number of students absent corresponds to the $\frac{1}{2}(n + 1)$th value.

n is the total frequency $= 5 + 14 + 12 + 4 + 2 + 1 + 1 = 39$

So the median is the $\frac{1}{2}(39 + 1)$th $=$ 20th value.

In order to find the 20th value, you need to find the running totals.

Number of students absent, x	Number of weeks, f	Running total
0	5	5
1	14	5 + 14 = 19
2	12	5 + 14 + 12 = 31
3	4	5 + 14 + 12 + 4 = 35
4	2	5 + 14 + 12 + 4 + 2 = 37
5	1	5 + 14 + 12 + 4 + 2 + 1 = 38
6	1	5 + 14 + 12 + 4 + 2 + 1 + 1 = 39

The 20th value lies in here. This corresponds to 2 students absent. This is the median.

The median value is 2 students absent.

c First add an extra column at the end of the table to contain the frequency × value (fx).

Number of students absent, x	Number of weeks, f	$f \times x$
0	5	$5 \times 0 = 0$
1	14	$14 \times 1 = 14$
2	12	$12 \times 2 = 24$
3	4	$4 \times 3 = 12$
4	2	$2 \times 4 = 8$
5	1	$1 \times 5 = 5$
6	1	$1 \times 6 = 6$
Totals	$\Sigma f = 39$	$\Sigma fx = 69$

AQA Examiner's tip

Make an extra column at the end of your table to put your calculations in. This will help the examiner to see your method.

The mean number of students absent per lesson is

$$\text{mean} = \frac{\text{the total of (frequencies} \times \text{values)}}{\text{the total of frequencies}} = \frac{\Sigma fx}{\Sigma f} = \frac{69}{39} = 1.7692...$$

$$\text{mean} = 1.77 \text{ (to 2 d.p.)}$$

d There are 32 students in the class and the mean number of students absent per lesson is 1.77
The mean number of students present must be $32 - 1.77 = 30.23$ students.

e The range is the difference between the highest and lowest values.
The highest number of students absent is 6.
The lowest number of students absent is 0.
The range is $6 - 0 = 6$ students.

D C

2 A school is raising money for a local charity.
The table below shows the mean amount of money raised per student in two classes.

Class	Numer of pupils	Mean amount raised
Mr Singh	25	£5.60
Mrs Johnson	18	£7.25

Find the total amount of money raised by the two classes.

Solution

2 To find the amount raised by each class, you need to multiply the mean amount raised by the number of students in the class.
Amount raised by Mr Singh's class is 25 × £5.60 = £140
Amount raised by Mrs Johnson's class is 18 × £7.25 = £130.50
Total amount raised is £140 + £130.50 = £270.50

Example Grouped frequency distributions Unit 1

C

The table shows the prices of 300 selected houses in a particular town during 2000 and 2008.

a For the year 2000, calculate the class interval in which the median house price lies.

b Find an estimate of the percentage increase in the mean price of the houses between 2000 and 2008.
Comment on the reliability of your answer.

House price, x, (£000s)	Number of houses in 2000	Number of houses in 2008
$60 \leqslant x < 100$	78	11
$100 \leqslant x < 150$	127	75
$150 \leqslant x < 200$	49	108
$200 \leqslant x < 300$	27	66
$300 \leqslant x < 400$	19	40

Solution

a The median class interval is the one that contains the $\frac{1}{2}(n + 1)$th value.
n is the total frequency, which is 300 houses.
So the median is the $\frac{1}{2}(300 + 1)$th $= 150.5$th value.
In order to find the 105.5th value you need to find the running totals.

House price, x, (£000s)	Number of houses in 2000	Running total
$60 \leqslant x < 100$	78	78
$100 \leqslant x < 150$	127	$78 + 127 = 205$
$150 \leqslant x < 200$	49	$78 + 127 + 49 = 254$
$200 \leqslant x < 300$	27	$78 + 127 + 49 + 27 = 281$
$300 \leqslant x < 400$	19	$78 + 127 + 49 + 27 + 19 = 300$

The 150.5th value lies in here. This corresponds to the class interval $100 \leqslant x < 150$

So the interval that contains the median is $100 \leqslant x < 150$

b In order to find an estimate of the percentage increase in the house prices between 2000 and 2008, you will first need to find an estimate of the mean for each of these years.

In order to find the mean, you need to complete the following calculation.

$$\text{mean} = \frac{\text{the total of (frequencies} \times \text{midpoint)}}{\text{the total of frequencies}} = \frac{\Sigma fx}{\Sigma f}$$

Add two extra columns at the end of the table.

In the first column, list the set of midpoints of the class intervals.

Use the second column to calculate frequency × midpoint.

House prices in 2000

House price, x (£000s)	f	Midpoint	Frequency × midpoint
$60 \leqslant x < 100$	78	80	$78 \times 80 = 6240$
$100 \leqslant x < 150$	127	125	$127 \times 125 = 15\,875$
$150 \leqslant x < 200$	49	175	$49 \times 175 = 8575$
$200 \leqslant x < 300$	27	250	$27 \times 250 = 6750$
$300 \leqslant x < 400$	19	350	$19 \times 350 = 6650$
Totals	$\Sigma f = 300$		$\Sigma fx = 44\,090$

House prices in 2008

House price, x (£000s)	f	Midpoint	Frequency × midpoint
$60 \leqslant x < 100$	11	80	$11 \times 80 = 880$
$100 \leqslant x < 150$	75	125	$75 \times 125 = 9375$
$150 \leqslant x < 200$	108	175	$108 \times 175 = 18\,900$
$200 \leqslant x < 300$	66	250	$66 \times 250 = 16\,500$
$300 \leqslant x < 400$	40	350	$40 \times 350 = 14\,000$
Totals	$\Sigma f = 300$		$\Sigma fx = 59\,655$

The mean house price in 2000 is

$$\text{mean} = \frac{\text{the total of (frequencies} \times \text{midpoint)}}{\text{the total of frequencies}} = \frac{\Sigma fx}{\Sigma f} = \frac{44\,090}{300} = 146.9666\ldots$$

As the house prices are given in £000s, an estimate of the mean house price is £146 967.

In 2008 an estimate of the mean house price is $\frac{59\,655}{300} = 198.85$, which is £198 850.

The percentage increase is given by

$$\% \text{ increase} = \frac{\text{increase}}{\text{original amount}} \times 100$$

$$= \frac{198\,850 - 146\,967}{146\,967} \times 100$$

$$= 35.3\%$$

This is only an estimate of the house price increase as you don't have the actual values.

Bump up your grade

To get a Grade C, you need to be able to find an estimate of the mean of data in a grouped frequency distribution.

Practise... Statistical measures Unit 1

D C B

D

1 Richard is trying to find the mean of the following frequency distribution.

x	Frequency
3	11
4	15
5	27
6	38

He says that the mean is 25.3...

a How do you know from the table that he is wrong?

b Find the mean of the data in the table.

D B

2 The number of letters received per day for 55 days is recorded.

Number of letters	Number of days
0	6
1	19
2	13
3	9
4	6
5	0
6	0
7	2

a Write down the modal number of letters received per day.

b Find the median number of letters received per day.

c Find the mean number of letters received per day.

d Find the range of letters received.

e Find the inter-quartile range of letters received.

f Explain why the inter-quartile range may be more useful than the range for these data.

C

3 Last year a European city decided to begin charging motorists to enter the city centre.

One year later the city authorities decided to look at the effect of the project.

The table below shows the number of occupants of all the cars entering the city between 8 am and 9 am on a typical work day.

Number of occupants	1	2	3	4	5	6
Number of cars before charging	658	275	86	25	4	1
Number of cars after charging	450	388	125	47	8	1

a Find the mean number of occupants per car before the charge was introduced.

b Find the mean number of occupants per car after the charge was introduced.

c Comment on the effect you think the charge has had.

4 A local health authority carried out a survey.

People were asked how many portions of fruit or vegetables they ate in one day.

The authority then carried out an 'Eat 5 a Day' campaign.

The same people were asked the same question one month after the campaign.

The results are shown in the table.

Has the advertising campaign been successful?

You need to back up your comments with calculations.

Number of portions	Before campaign	After campaign
0	57	45
1	1210	954
2	2123	1579
3	7876	6650
4	6601	7332
5	1955	2887
6	604	950
7	70	88
8	9	20

5 The table below shows the speed of 80 vehicles on a motorway.

Speed (mph)	$50 \leqslant x < 60$	$60 \leqslant x < 70$	$70 \leqslant x < 80$	$80 \leqslant x < 90$	$90 \leqslant x < 100$
Frequency	4	38	29	6	3

a A motoring group claims that on average car drivers travel between 60 to 70 mph, and therefore are not breaking the speed limit. Which average is the motoring group using here?

b A road safety group claims that on average car drivers **are** breaking the speed limit. Use an estimate of the mean speed to show that the road safety group is correct.

6 The mean number of marks that 22 students obtained on a test was 26.5.

Two more students did the test at a later date.

They received 35 and 28 marks on the test.

Find the new mean number of marks obtained by all 24 students.

Give your answer correct to one decimal place.

B

7 The number of passengers in 100 cars is recorded.

Estimate what percentage of cars carried more than the mean number of passengers.

Number of passengers	Number of cars
0	34
1	29
2	21
3	10
4	4
5	2

Statistics

3

Representing data

Unit 1

7 Representing data
8 Scatter graphs

Key terms

Write down definitions for the following words. Check your answers in the glossary of your Student Book.

- back-to-back stem-and-leaf diagram
- coordinate
- correlation
- cumulative frequency
- cumulative frequency diagram
- frequency diagram
- frequency polygon
- histogram
- inter-quartile range
- line graph
- line of best fit
- lower quartile
- negative correlation
- outlier
- positive correlation
- scatter graph
- stem-and-leaf diagram
- upper quartile
- zero or no correlation

Revise... Key points

This chapter looks at using diagrams to represent and interpret data. Representing data is part of the third stage of the data handling cycle. The fourth stage is to interpret the diagrams.

Stem-and-leaf diagram (Unit 1)

A **stem-and-leaf diagram** is a way of representing data.

A stem-and-leaf diagram allows you to see the original data values.

Stem-and-leaf diagrams require a key so that you can interpret the values.

Stem-and-leaf diagram to show the heights of students on a minibus

Stem	Leaves
15	2
16	3 4 7 7
17	0 2 3 5 8
18	1 1
19	0

stem → 17; leaves → 8, 1, 0

Key: 16 | 3 represents 163 cm

Each leaf represents one value. So in this example, the heights of 13 students were measured.

Two data sets can be shown at the same time on a **back-to-back stem-and-leaf diagram**.

Link

See Chapter 7 in the Student Book, Learn 7.1, for an example of a back-to-back stem-and-leaf diagram.

AQA Examiner's tip

In an exam you may be asked to draw a stem-and-leaf diagram.
Ensure that you draw an ordered stem-and-leaf diagram.
This means you need to put the numbers in ascending order.

Don't forget to include a key for your diagram.

Line graphs and frequency polygons (Unit 1)

A **line graph** is a series of points joined with straight lines.

Line graphs show how data changes over a period of time.

A **frequency polygon** is an example of a **frequency diagram**.

It is used to display continuous grouped data.

For a frequency polygon, you plot the frequency at the midpoint of the class interval.

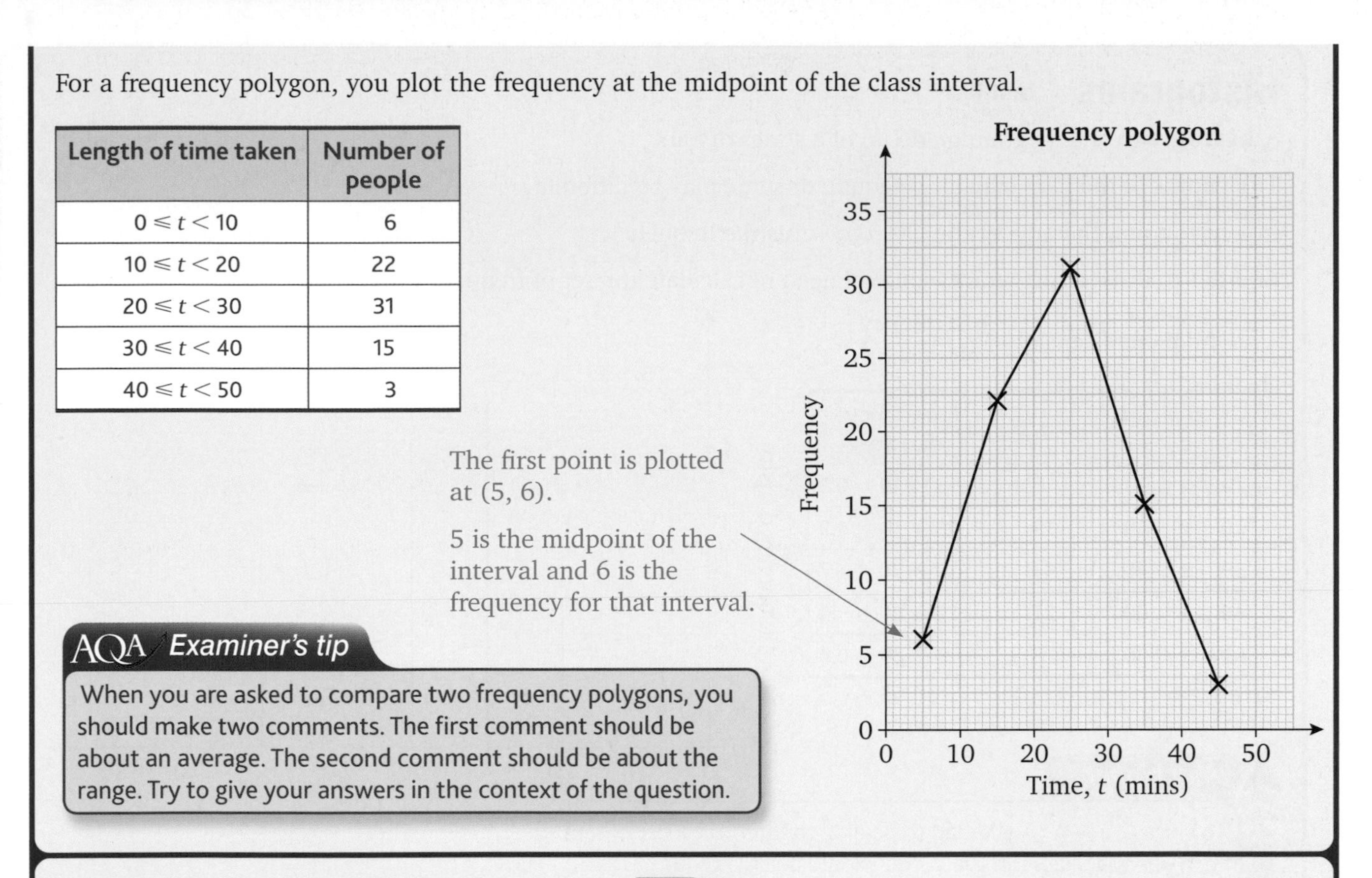

Length of time taken	Number of people
$0 \leq t < 10$	6
$10 \leq t < 20$	22
$20 \leq t < 30$	31
$30 \leq t < 40$	15
$40 \leq t < 50$	3

The first point is plotted at (5, 6).

5 is the midpoint of the interval and 6 is the frequency for that interval.

AQA Examiner's tip

When you are asked to compare two frequency polygons, you should make two comments. The first comment should be about an average. The second comment should be about the range. Try to give your answers in the context of the question.

Cumulative frequency diagram Unit 1

Cumulative frequency is a running total of the frequency.

A **cumulative frequency diagram** is used to estimate the median, lower quartile and upper quartile from a set of data.

You plot the cumulative frequency at the upper class boundary of the class interval.

Points are either joined by a series of straight lines (polygon), or you may use a smooth curve.

The **inter-quartile range** tells you the range of the middle 50% of the data.

It is calculated using

inter-quartile range = upper quartile − lower quartile

Box plots Unit 1

A box plot shows a graphical summary of the data. It shows the following information.

- the minimum and maximum values
- the lower and upper quartiles
- the median.

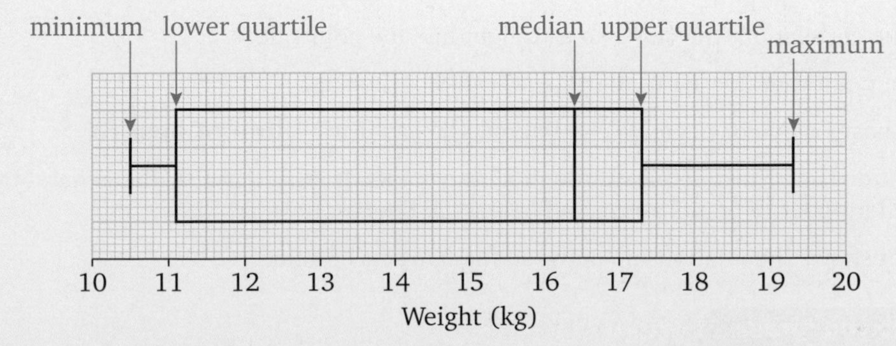

To find the range, you find the difference between the highest and lowest values.

To find the inter-quartile range, you find the difference between the upper quartile and lower quartile.

Histograms Unit 1

A **histogram** is a diagram made up of a series of bars.

The widths of the bars may all be equal, or some may be different.

In a histogram, the area of the bar represents the frequency.

In order to draw a histogram, you will need to calculate the set of frequency densities.

$$\text{frequency density} = \frac{\text{frequency}}{\text{class width}}$$

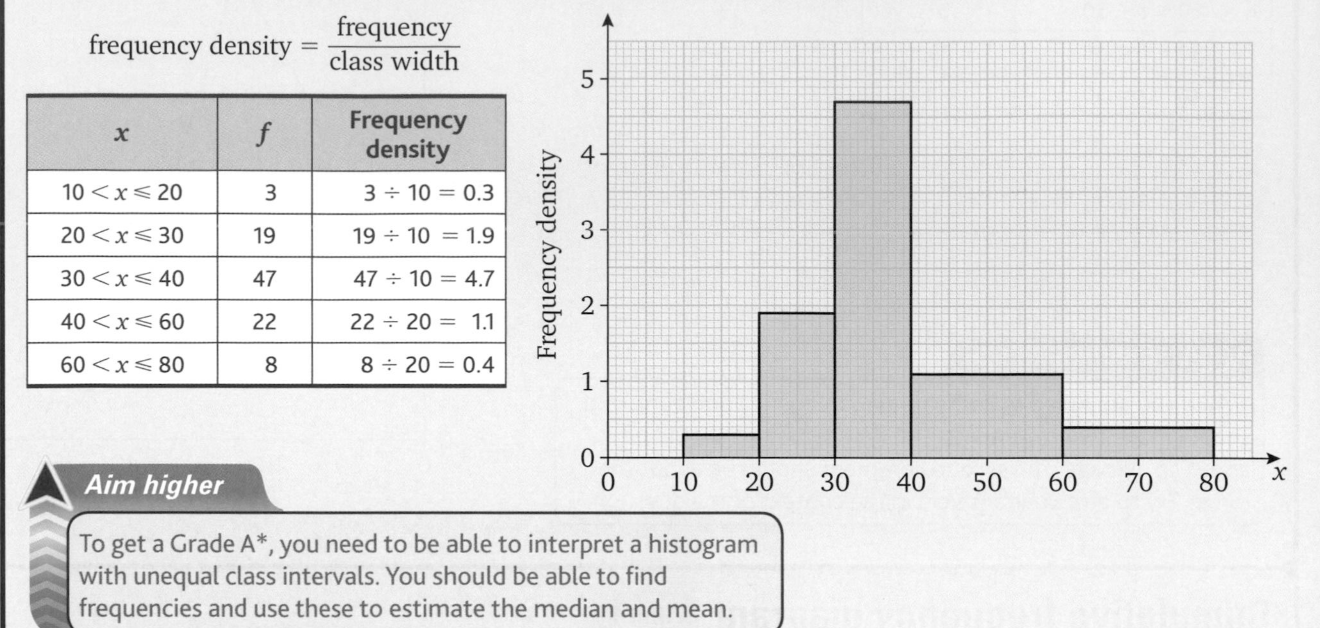

x	f	Frequency density
$10 < x \leqslant 20$	3	$3 \div 10 = 0.3$
$20 < x \leqslant 30$	19	$19 \div 10 = 1.9$
$30 < x \leqslant 40$	47	$47 \div 10 = 4.7$
$40 < x \leqslant 60$	22	$22 \div 20 = 1.1$
$60 < x \leqslant 80$	8	$8 \div 20 = 0.4$

Aim higher

To get a Grade A*, you need to be able to interpret a histogram with unequal class intervals. You should be able to find frequencies and use these to estimate the median and mean.

Scatter graphs Unit 1

Scatter graphs are used to show the relationship between two sets of data.

Correlation measures the relationship between two sets of data.

There are three types of correlation that you are likely to meet.

Positive correlation

(as one variable increases the other increases)

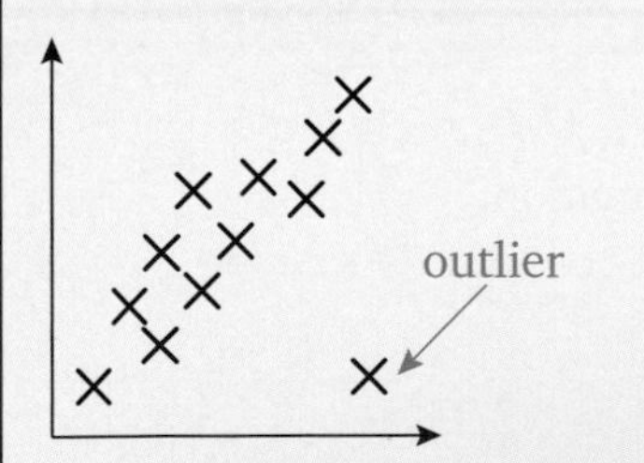

Negative correlation

(as one variable increases the other decreases)

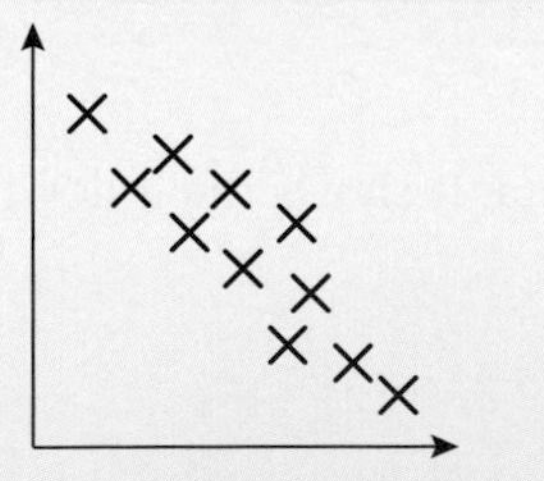

Zero or no correlation

(points are scattered all over the diagram)

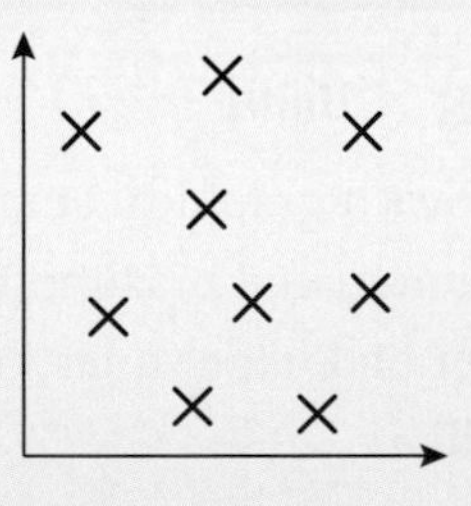

The stronger the correlation, the closer to a straight line the points lie.

An **outlier** is a point that doesn't fit the general trend of the data.

When a graph has positive or negative correlation, you can draw a **line of best fit**.

A line of best fit does not have to pass through all the points. It should follow the general trend of the points, with a balanced number of points above and below the line.

A line of best fit shows the relationship between the two sets of data.

Bump up your grade

To get a Grade C, you should be able to draw a line of best fit on a scatter graph and use your line to estimate values.

Example Stem-and-leaf diagrams Unit 1

D

The stem-and-leaf diagram shows the number of marks 25 students scored in an IQ test.

a Find the median score.

b Find the range of scores.

c What percentage of students scored more than 30 marks?

Stem	Leaves
1	5 6 6 7
2	2 3 4 5 5 8
3	0 1 4 7 9 9 9
4	2 5 6 7
5	1 2 3
6	0

Key: 2 | 3 represents a score of 23 marks

Solution

a The median score is the $\frac{1}{2}(n + 1)$th value. (Where n is the number of values.)

The median is the $\frac{1}{2}(25 + 1) = 13$th value.

The 13th value from the stem-and-leaf diagram is 34.

b The highest value is 60. The lowest value is 15.

The range is the difference between the highest and lowest values $= 60 - 15 = 45$ marks.

c There are 14 students who scored more than 30 marks.

25 students took the IQ test in total.

The percentage of students who scored more than 30 marks is $\frac{14}{25} \times 100 = 56\%$

Example Cumulative frequency diagrams/box plots Unit 1

B

A consumer magazine is comparing the lifetimes of two types of battery, A and B.

The lifetime in hours of 200 batteries of Brand A is shown in the table below.

Lifetime (hours)	Number of batteries
$100 < w \leq 120$	8
$120 < w \leq 140$	58
$140 < w \leq 160$	122
$160 < w \leq 180$	7
$180 < w \leq 200$	5

a Draw a cumulative frequency diagram to show this information.

The box plot below shows the lifetime of 200 batteries for Brand B.

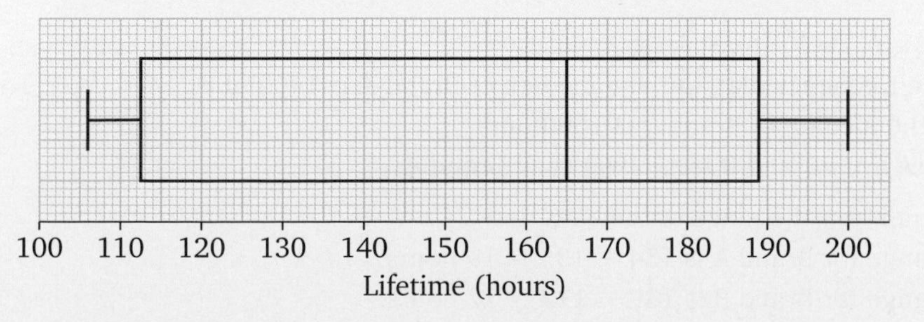

b Compare the lifetimes of the two batteries.

You should show any calculations that you have made.

Solution

a First find the set of cumulative frequencies.

Lifetime (hours)	Number of batteries	Cumulative frequency
$100 < w \leqslant 120$	8	8
$120 < w \leqslant 140$	58	$8 + 58 = 66$
$140 < w \leqslant 160$	122	$66 + 122 = 188$
$160 < w \leqslant 180$	7	$188 + 7 = 195$
$180 < w \leqslant 200$	5	$195 + 5 = 200$

Cumulative frequency is a running total of the frequency. Simply add on the next frequency value.

AQA Examiner's tip

Check that your final cumulative frequency value matches the information given in the question (for example, this question featured 200 batteries).

Now plot the cumulative frequencies at the end point of the class intervals.

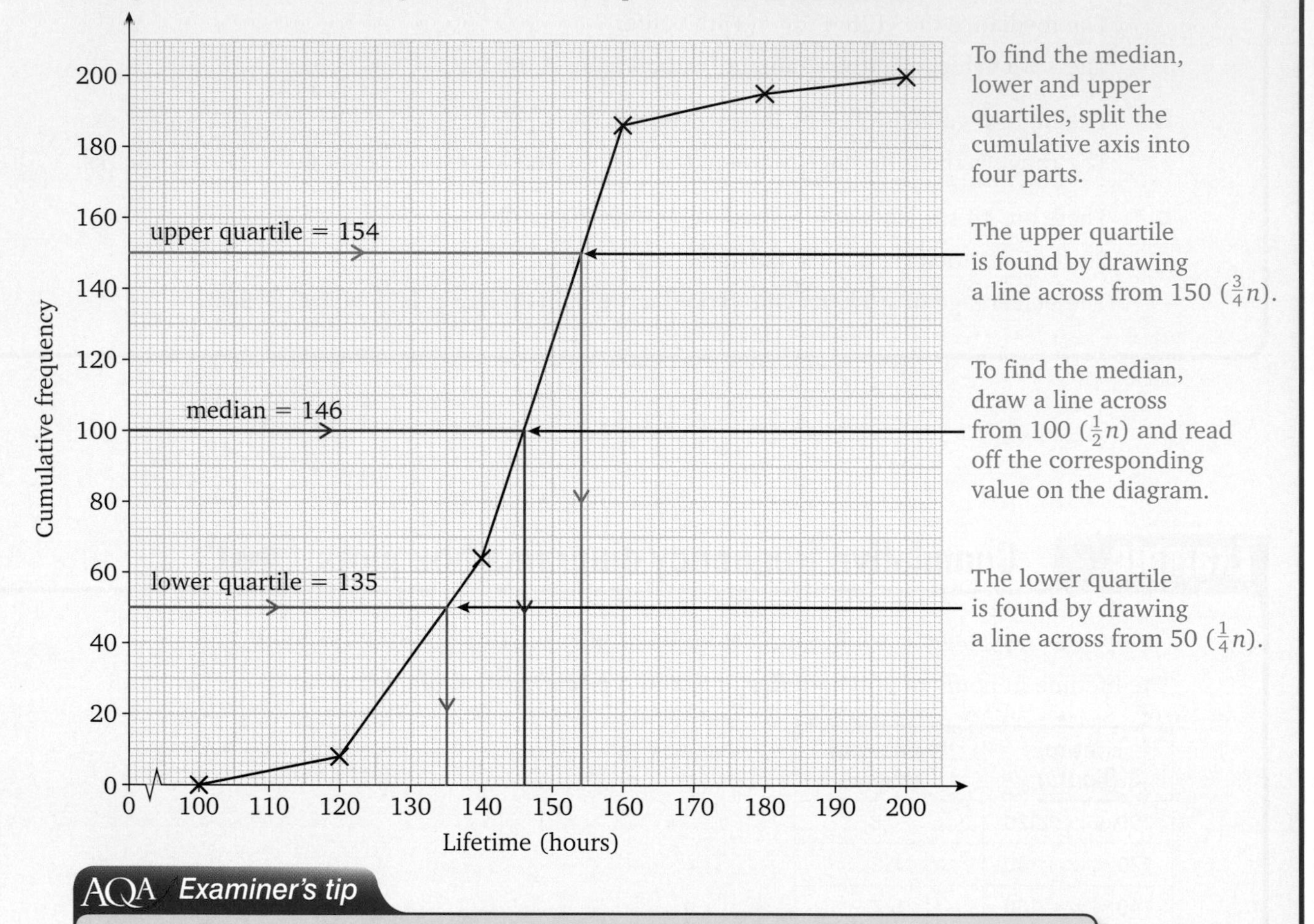

AQA Examiner's tip

Your cumulative frequency diagram should always be increasing. If it isn't, you have probably plotted the frequency values by mistake.

On a cumulative frequency diagram the median and quartiles are read off at $\frac{1}{2}n$, $\frac{1}{4}n$ and $\frac{3}{4}n$.

b To answer this question, you should make two comparisons.

You should make one comparison about the average and one comparison about the range or inter-quartile range.

First make a comparison about the average.

From the cumulative frequency diagram, the median of Brand A is 146 hours.

From the box plot, the median of Brand B is 165 hours.

On average, batteries from Brand B last longer than Brand A.

Now compare the inter-quartile range of the batteries.

The inter-quartile range for Brand A is $154 - 135 = 19$ hours.

The inter-quartile range for Brand B is $189 - 112 = 77$ hours.

This means that Brand A batteries are more consistent and their lifetimes do not vary as much as Brand B batteries.

Example Histograms Unit 1

A*

The histogram shows the amount of time some students spent on a piece of GCSE art homework.

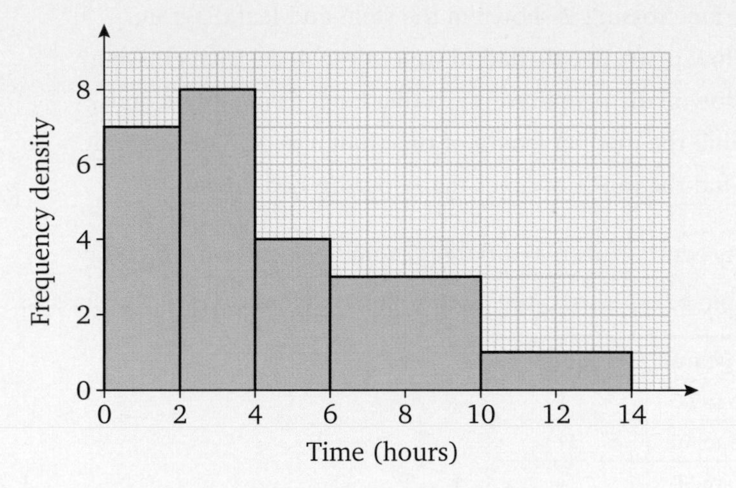

Estimate the mean amount of time spent on this piece of homework.

Solution

In order to find the mean time, you first need to calculate the frequencies.

Number of hours	Frequency (frequency density × class width)
0–2	7 × 2 = 14
2–4	8 × 2 = 16
4–6	4 × 2 = 8
6–10	3 × 4 = 12
10–14	1 × 4 = 4

The first bar has a height of 7 units (frequency density) and is 2 units wide (class width). Multiply these to get the frequency.

Hint

If a histogram is already given, you can find the frequencies using

frequency = frequency density × class width

AQA Examiner's tip

Organise your frequencies into a table. This will help you with further calculations.

In order to find an estimate of the mean we use

$$\text{estimate of mean} = \frac{\text{the total of (frequencies} \times \text{midpoint)}}{\text{the total of frequencies}} = \frac{\Sigma fx}{\Sigma f} = \frac{246}{54} = 4.6 \text{ hours (1 d.p.)}$$

Link

See Chapter 2 Statistical measures for how to find an estimate of the mean of data in a table.

Practise... Representing data Unit 1

D C B A A*

D

1 The length of time that some people have to wait at a pedestrian crossing is shown in the stem-and-leaf diagram.

0	4 8 9
1	2 2 5 7 8
2	3 5 6
3	5 8
4	2 3 3

Key: 1 | 2 represents 12 seconds

a How many people had to wait at the pedestrian crossing?

b How many people were waiting longer than 30 seconds?

c Find the median length of time that a person had to wait.

d Find the range of times that a person had to wait.

D C

2 A sports centre runs two different exercise classes on a Monday.

The table below shows the distribution of ages at one of the classes.

Age (years)	Frequency
20 up to 30	5
30 up to 40	11
40 up to 50	24
50 up to 60	36
60 up to 70	8

a Draw a frequency polygon for these data.

The frequency polygon below shows the distribution of ages at the second class.

b Which exercise class is more popular?

c Compare the ages of people who attend the two classes.

C

3 In a science experiment, the amount of salt dissolving, in milligrams, was measured every one minute.

The results are shown in the table below.

Time (minutes)	1	2	3	4	5	6	7	8	9	10
Amount of salt remaining (mg)	40	38	35	34	32	30	29	28	25	25

a Plot this information on a scatter diagram and draw a line of best fit.

b Use your line of best fit to estimate how many milligrams of salt will remain after 14 minutes.

c How reliable is your estimate? Explain your answer.

AQA Examiner's tip

When drawing a line of best fit, make sure you follow the general trend of the points and leave about the same number of points either side of the line. A line of best fit does not have to pass through the origin.

B

4 The cumulative frequency diagram shows the weights of 500 pies.

How many pies weighed between 70 g and 125 g?

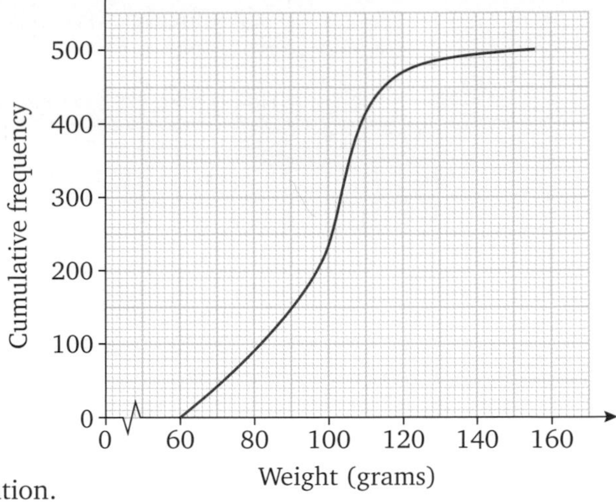

A

5 Draw a histogram to show the following information.

Weight (kg)	Frequency
$0 < w \leqslant 2$	8
$2 < w \leqslant 4$	17
$4 < w \leqslant 10$	42
$10 < w \leqslant 15$	26
$15 < w \leqslant 25$	15

A*

6 The histogram below shows the height, in centimetres, of a sample of flowers.

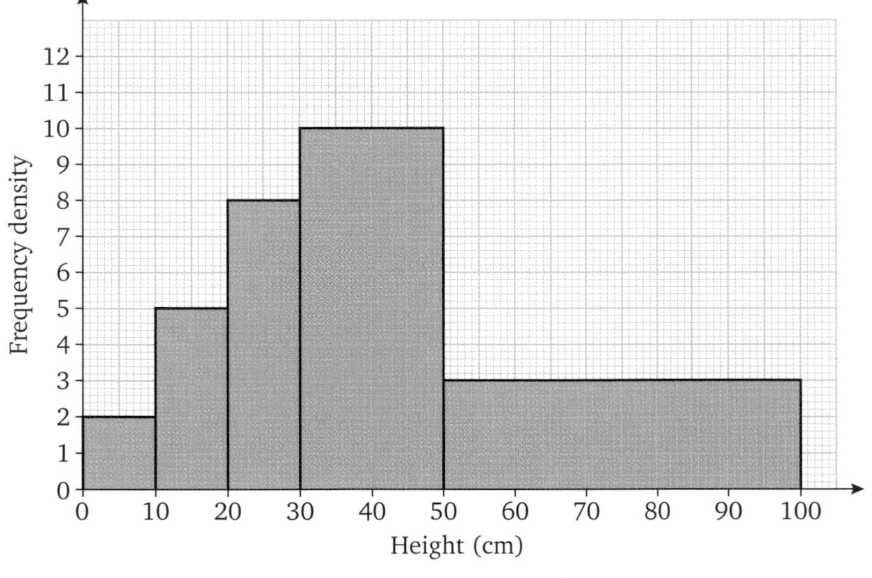

a How many flowers have a height greater than 75 cm?

b Calculate an estimate for the median height of the flowers.

7 The histogram shows the distribution of marks on a French test.

15% of students achieved a Grade A*.

Estimate the mark needed in order to achieve an A*.

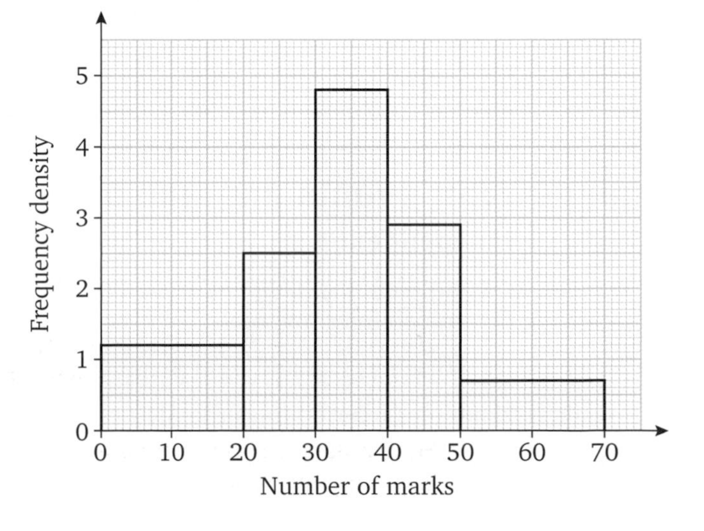

Statistics

4 Probability

Unit 1

9 Probability

Key terms

Write down definitions for the following words. Check your answers in the glossary of your Student Book.

biased
conditional probability
dependent events
experimental probability
fair
independent events
mutually exclusive events
random
relative frequency
theoretical probability
tree diagram
trial
two-way table

AQA *Examiner's tip*

Try to always use fractions or decimals to write probabilities. Never write a probability as a ratio. You will gain no marks for this in the exam.

Only cancel down a fraction if it asks you to give your answer in simplest form.

Revise... Key points

Mutually exclusive events 

You should try to use shorthand probability notation.

The probability of event A happening can be written as P(A).

Mutually exclusive events are events that cannot happen at the same time.

If two events A and B are mutually exclusive then

$$P(A \text{ or } B) = P(A) + P(B)$$

This rule is sometimes referred to as the OR rule in probability.

If the mutually exclusive events listed cover all the possibilities, the sum of their probabilities must equal 1.

Relative frequency Unit 1

Relative frequency is also known as **experimental probability**.

The relative frequency of an event is the probability of the event happening based on the results of an experiment.

$$\text{relative frequency of an event} = \frac{\text{number of times an event has happened}}{\text{total number of trials}}$$

Relative frequency can also be based on past experience.

You can use results of an experiment to estimate the probability of an event.

The more times you carry out an experiment, the more reliable your probability estimate will be.

To work out the expected number of times a particular event will happen, you should multiply the probability by the number of times you are going to carry out the experiment.

Bump up your grade

In order to get a Grade C, you need to be able to calculate relative frequencies and plot them on a graph.

Independent events and tree diagrams 

Events are **independent** if the outcome of one event does affect the outcome of the other.

If two events A and B are independent, the probability that event A and event B happen is found by multiplying the probabilities together.

It is sometimes written as $P(A \text{ and } B) = P(A) \times P(B)$

This rule is sometimes referred to as the AND rule for probability.

A **tree diagram** is a useful way of finding probabilities of more than one event happening.

A tree diagram is made up of branches.
At the end of each branch you write the event.

The probability of the event happening is written on the branch.

Probabilities on pairs of branches must add up to 1.

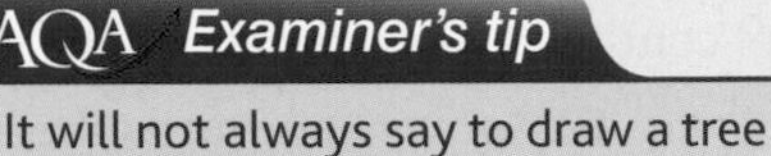

AQA Examiner's tip

It will not always say to draw a tree diagram in an exam question. It will usually help to draw a tree diagram when there are two or more events happening.

Dependent events and conditional probability Unit 1

Events are **dependent** if the outcome of one event is affected by the outcome of another.

Once again, a tree diagram is usually the most useful way of answering this type of question.

AQA Examiner's tip

Often the words 'without replacement' are used. This means that after the first item has been removed it is not placed back.

Aim higher

In order to get a Grade A/A*, you need to be able to calculate probabilities of events happening where an item has not been replaced.

Example Mutually exclusive events Unit 1

D

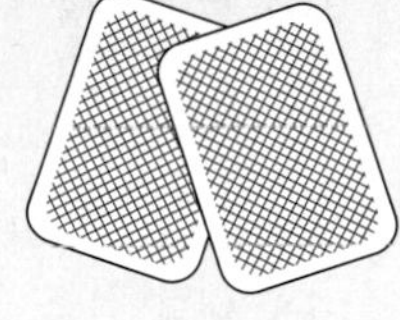

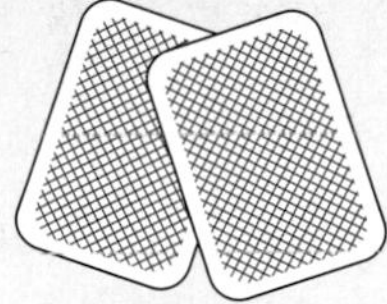

A game involves choosing a card at random from 250 cards that are laid face down.

All the cards are numbered either 1, 2, 3 or 4.

The probability of choosing a particular numbered card is shown in the table below.

Card number	1	2	3	4
Probability	0.2	0.12		

There are twice as many cards numbered 3 as there are numbered 2.

Use all this information to work out how many cards are numbered 4.

Solution

There are twice as many cards numbered 3 as there are numbered 2.

Therefore, the probability of getting a 3 must be twice the probability of getting a 2.

Probability of getting a card numbered 3 $= 0.12 \times 2 = 0.24$

The total probability in the table must equal 1 (as the events are mutually exclusive and they are all listed).

So the probability of getting a card numbered 4 must be

$$1 - (0.2 + 0.12 + 0.24) = 0.44$$

To work out the number of cards numbered 4 you need to multiply the probability by the total number of cards.

So number of cards numbered 4 is $0.44 \times 250 = 110$ cards.

Example Relative frequency Unit 1

C

Daniel is carrying out an experiment to test whether a coin is **biased**.

He throws the coin 10 times and obtains 7 heads.

a Do these results suggest the coin is biased?

Explain your answer.

C

Daniel continues to throw the coin.

After every 10 throws he records the total number of heads obtained so far.

The results are shown in the table below.

Number of throws	10	20	30	40	50	60	70	80	90	100
Total number of heads	7	16	25	29	35	44	52	59	66	74

b Plot the results on a relative frequency graph.

In fact the coin is biased.

c Write down an estimate of the probability of the coin landing on heads.

Solution

a No. Although the coin lands on heads more times than expected, Daniel has only thrown the coin 10 times. It is not possible to tell whether the coin is biased from such a small number of trials.

Daniel needs to throw the coin more times.

b You first need to work out the relative frequencies.

For 10 throws the relative frequency will be $\frac{7}{10} = 0.7$

For 20 throws the relative frequency will be $\frac{16}{20} = 0.8$

You can work out the other relative frequencies in the same way.

Number of throws	10	20	30	40	50	60	70	80	90	100
Total number of heads	7	16	25	29	35	44	52	59	66	74
Relative frequency	$\frac{7}{10} = 0.7$	$\frac{16}{20} = 0.8$	$\frac{25}{30} = 0.83$	$\frac{29}{40} = 0.73$	$\frac{35}{50} = 0.7$	$\frac{44}{60} = 0.73$	$\frac{52}{70} = 0.74$	$\frac{59}{80} = 0.74$	$\frac{66}{90} = 0.73$	$\frac{74}{100} = 0.74$

You can now plot these points on a relative frequency graph.

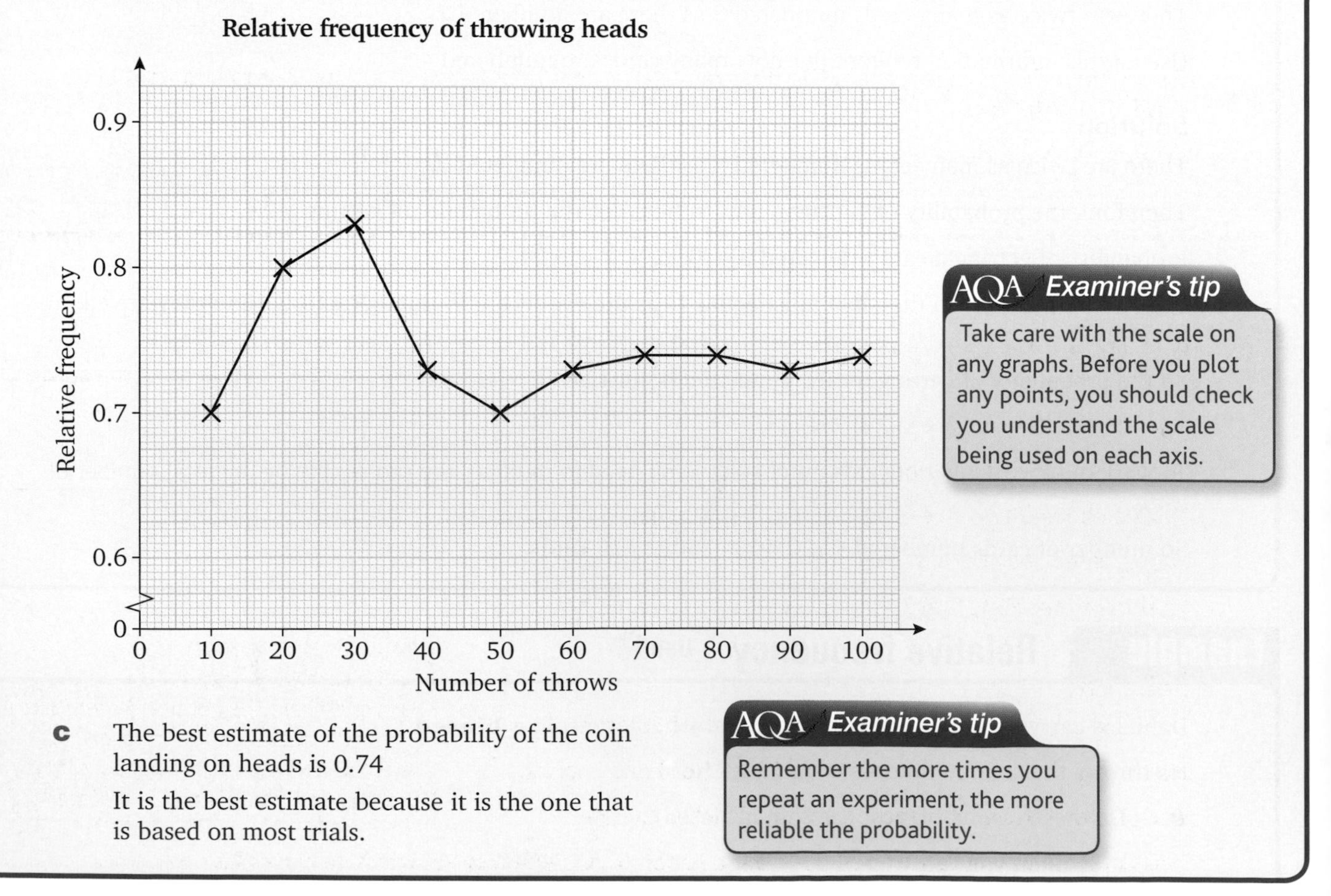

AQA Examiner's tip

Take care with the scale on any graphs. Before you plot any points, you should check you understand the scale being used on each axis.

c The best estimate of the probability of the coin landing on heads is 0.74

It is the best estimate because it is the one that is based on most trials.

AQA Examiner's tip

Remember the more times you repeat an experiment, the more reliable the probability.

Example Independent events and tree diagrams Unit 1

B

The probability that Charley goes to the gym on Monday is 0.8

The probability that Charley goes to the gym on Tuesday is 0.6

What is the probability that Charley goes to the gym on either Monday or Tuesday, but not both?

Solution

To answer this question, you could draw a tree diagram.

AQA Examiner's tip

Think carefully about how to label the tree diagram before you start. A common mistake would be to label the first pair of branches as Monday and Tuesday.

Check that the probabilities on pairs of branches all add up to one.

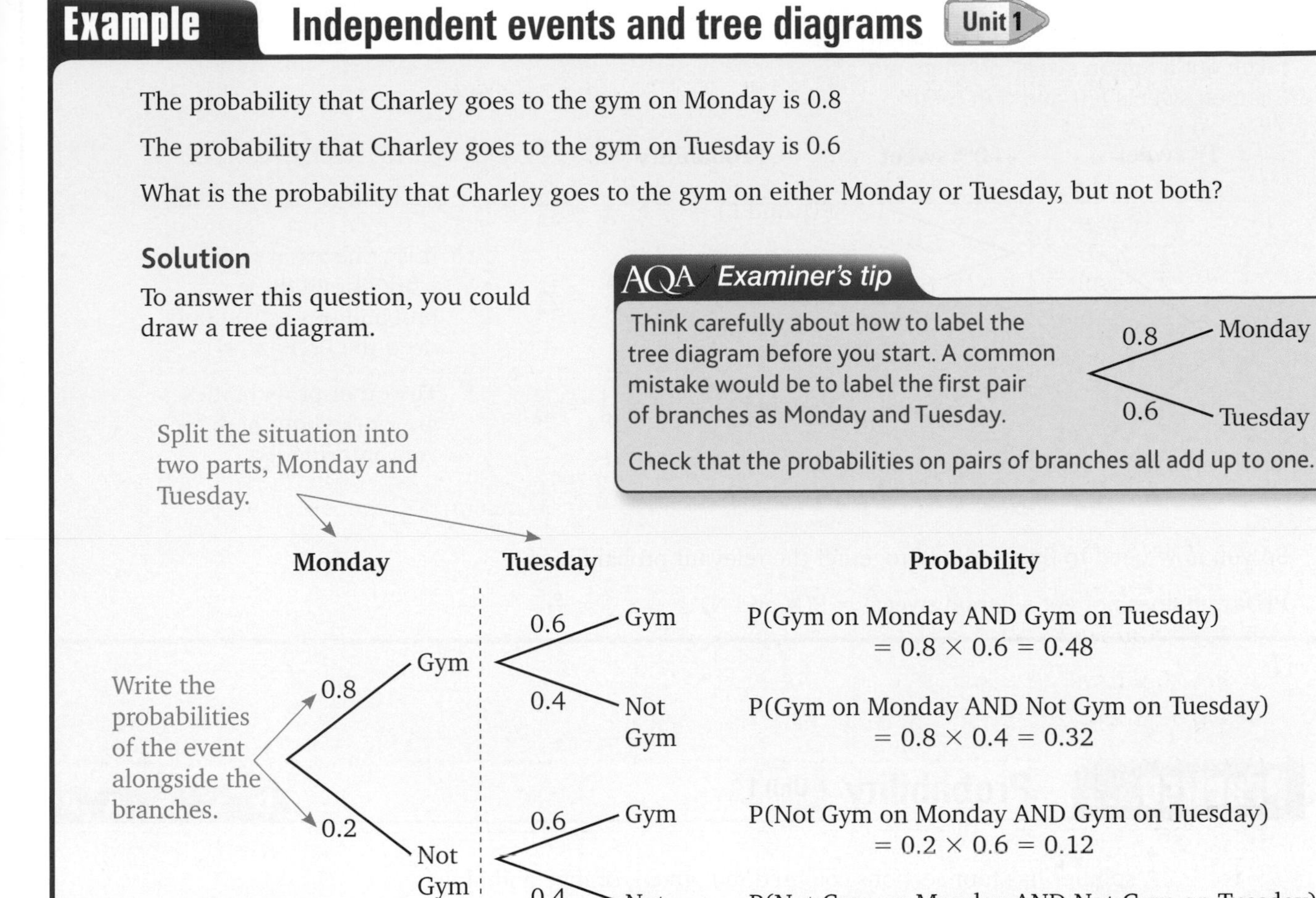

To work out the probability that Charley goes to the gym on only one day, you need to add up the correct probabilities.

P(Gym on either Monday or Tuesday, but not both) $= 0.8 \times 0.4 + 0.2 \times 0.6$

$= 0.32 + 0.12 = 0.44$

Example Dependent events and conditional probability Unit 1

A*

A jar contains some sweets.

Four of the sweets are lemon and the other three are not lemon.

Darren chooses two sweets at random from the jar.

Darren does not like lemon sweets.

Find the probability that Darren does not get a lemon sweet.

Solution

To answer this question, you could draw a tree diagram.

In the diagram L = Lemon sweet and N = Not lemon sweet.

The probabilities on the second set of branches change.

This depends on whether Darren takes out a lemon or not lemon sweet when he makes his first choice.

AQA Examiner's tip

This question does not use the words 'without replacement'. However, by selecting two sweets the implication is that the first one will not be replaced.

Here the probability is $\frac{3}{6}$ as Darren has taken out a lemon sweet, so there are 3 lemon sweets left and 6 in total.

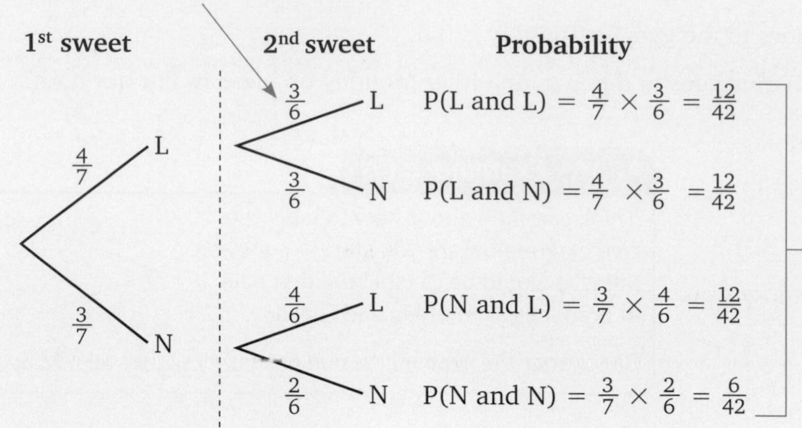

It isn't necessary to work out all these probabilities as you only need the P(N and N).

The other probabilities are worked out here for your information.

So you now need to find and add together the relevant probabilities.

P(Darren does not get a lemon sweet) = P(N and N) = $\frac{3}{7} \times \frac{2}{6} = \frac{6}{42}$

Practise... Probability Unit 1

D C B A A*

D

1 A spinner has four sections coloured red, green, orange and blue.

The table below shows the probability of spinning each colour.

Colour	Red	Green	Orange	Blue
Probability	0.3	0.4	0.2	

a What is the probability of spinning red or green?

b What is the probability of spinning blue?

c What is the probability of not spinning red?

d I spin the spinner 120 times.
How many times do I expect to spin red?

2 James, Andy and Dominic are playing card games.

The probability of each of them winning is shown in the table.

Player	James	Andy	Dominic
Probability	0.25	0.15	0.6

One evening they play 60 games.

How many more games would Dominic expect to win than Andy?

C

3 Julia is shooting arrows at an archery target.

Each time she aims for the centre.

She records the number of times she hits the centre of the target after every 10 attempts.

Number of attempts	10	20	30	40	50	60	70	80
Number of centre hits	3	4	10	11	13	16	17	21

a Find the relative frequency of hitting the centre after 10 attempts.

b Find the relative frequency of hitting the centre after 40 attempts.

c Draw a relative frequency polygon for this data.

d Write down an estimate of the probability that Julia hits the centre target on her next shot.

B

4 A box contains 10 coloured bean bags.
Three of the bean bags are green and seven are yellow.
A bean bag is chosen at random from the box and replaced.
Another bean bag is then selected.

a Use a tree diagram to find the probability that the two bean bags chosen are different colours.

b Calculate the probability that the two bean bags will be different colours if the first bean bag is not replaced.

A

5 Asif has five 10p coins, three 20p coins and two 50p coins in his moneybox.
He selects two coins at random from the moneybox.

a Find the probability that both coins are the same.

b Find the probability that the total of the coins is more than 50p.

c Find the probability that at least one of the coins is a 10p.

A*

6 The probability that Georgina oversleeps is 0.3
If Georgina oversleeps, the probability that she will miss the bus is 0.9
If she does not oversleep, the probability that she will miss the bus is 0.2
Find the probability that Georgina misses the bus.

7 A box contains 8 red balls, 5 blue balls and 2 green balls.
A charity game involves removing two balls from the box.
If you remove two balls of the same colour, you win.
The charity plans to charge £1 per game.
How much would you recommend the charity pays out if a player wins?
You should show all your working to justify your answer.

8 A board game involves moving around a board.

Below is a section of the board.

If a player lands on a picture of a dice, they roll again.

Player A is on square 24.

Player B is on square 19.

It is player B's turn. He rolls a fair six-sided dice.

What is the probability that player B overtakes player A?

5 AQA Examination-style questions

AQA Examination-style questions

D

1 A four-sided spinner is shown.

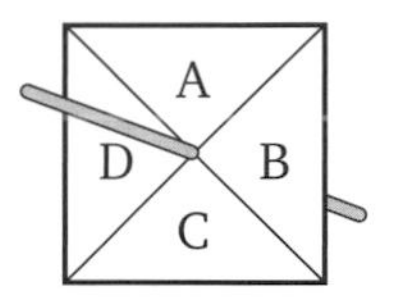

The spinner is biased.

Some of the probabilities of the spinner landing on a letter are shown in the table.

Letter	Probability
A	0.3
B	
C	
D	0.1

The probability that the spinner lands on *C* is double the probability that the spinner lands on *D*. The spinner is spun 60 times.

Calculate the number of times you would expect it to land on *B*. *(5 marks)*

AQA 2009

2 The table shows the number of bedrooms in each of 1000 houses.

Number of bedrooms	1	2	3	4	5	6
Number of houses	33	124	237	380	186	

The table is incomplete.

a How many houses have 6 bedrooms? *(1 mark)*

b State the modal number of bedrooms. *(1 mark)*

c Calculate two other averages for the number of bedrooms per house. *(4 marks)*

D C

3 A swimming pool is hosting a charity swim.

People are asked to swim as many lengths as they can in 30 minutes.

The results of 20 people who took part are shown below.

25	33	28	12	15	41	39	32	26	17
8	40	28	36	25	16	22	19	35	27

a Draw an ordered stem-and-leaf diagram to show this information. *(3 marks)*

b Another person swims 24 lengths.
Does this cause the median number of lengths to increase or decrease?
Remember to show your working. *(4 marks)*

D C

4 The table shows the area (in hectares) of 10 towns and their population.

Town	Area	Population
Grimsby	2300	87 000
Holmfirth	900	23 000
Lancaster	900	46 000
Longbenton	1000	35 000
Macclesfield	1200	27 000

Town	Area	Population
Northwich	1200	40 000
Otley	400	15 000
Richmond	200	8000
Rochdale	2200	96 000
Widnes	1600	56 000

a Draw a scatter diagram to represent this information. *(2 marks)*

b Describe the type and strength of the correlation. *(2 marks)*

c Draw a line of best fit on the scatter diagram. *(1 mark)*

d Approximately 29 000 people live in Shipley.
Use the line of best fit to estimate the area of Shipley. *(1 mark)*

C

5 A recent newspaper article made the following claim.

Boys receive nearly twice as much pocket money as girls

Mr Chadwick decides to carry out a survey with his class to check the claim.
The results of his survey are shown in the table below.

Amount of pocket money	Number of girls	Number of boys
$£0 < x \leq £3$	1	0
$£3 < x \leq £5$	18	7
$£5 < x \leq £10$	9	22
$£10 < x \leq £20$	2	1
£20+	0	0

Hint

Think about finding some averages and use them to make comparisons.

Do the results of Mr Chadwick's survey support the newspaper's claim? *(5 marks)*

6 A dice that is believed to be fair is rolled 30 times.

a How many times do you expect each number to appear? *(1 mark)*

The dice is thrown 30 times. The results are shown below.

2 2 3 3 5 3 3 6 1 3
5 3 3 3 3 2 4 1 5 4
6 2 3 3 2 3 1 4 3 3

b Copy and complete the relative frequency table below. *(2 marks)*

Score	1	2	3	4	5	6
Relative frequency						

c Do you think the dice is fair? Explain your answer. *(2 marks)*

B

7 A PE department wishes to compare girls' and boys' fitness levels in Year 7.
Write a brief report to explain how the PE department could investigate this.
Your answer should refer to the stages of the data handling cycle. *(4 marks)*

8 The diameters of 40 trees in a wood are measured.

The results are shown in the table.

Diameter (cm)	Frequency
5 up to 10	5
10 up to 15	12
15 up to 20	14
20 up to 25	7
25 up to 30	2

a Draw a cumulative frequency diagram to represent this data. *(3 marks)*

b Use the diagram to find:

i the median

ii the inter-quartile range. *(3 marks)*

c The smallest tree has a diameter of 6.5 cm.

Represent this information on a box plot. *(3 marks)*

The diameters of another set of trees in a different wood are measured.
The box plot shows the information gathered.

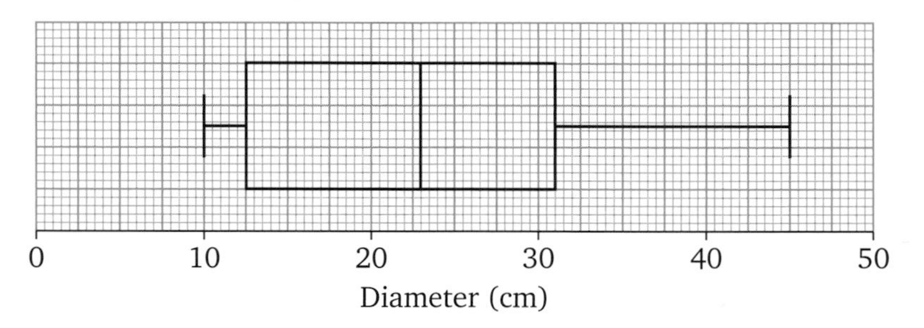

d Compare the trees in the two woods. *(2 marks)*

9 A fairground game consists of seven numbered discs.

Four of the discs are numbered 1, two are numbered 2 and the last disc is numbered 3.

The seven discs are placed into a bag.

A player chooses a disc at random.

The player then replaces the disc into the bag and takes out a second disc.

A player wins the game if they get a total of 4.

a Find the probability of winning the game. *(3 marks)*

b Find the probability of losing the game. *(1 mark)*

In another game, the disc is not replaced.

c Show that a player is less likely to win the game when the disc is not replaced. *(3 marks)*

B

10 The cumulative frequency diagram below shows the pulse rate of 50 people after exercise.

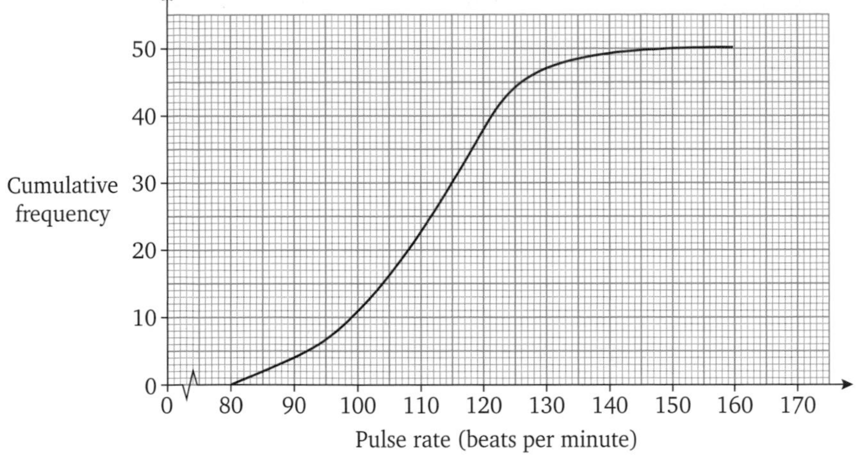

Hint
Create a table of frequencies for different class intervals (for example 80 to 90 etc…).

Find an estimate of the mean pulse rate.
Remember to show all the steps in your working. *(5 marks)*

A

11 The number of houses in a town is 12 000.
The number of different types of house is shown in the table below.

Bungalow	Terraced	Semi-detached	Detached
700	2500	7600	1200

An inspector wishes to visit 300 of the houses.
The inspector plans to use the method of stratified sampling.
How many houses of each type should she visit? *(4 marks)*

A A*

12 a The lengths of calls at a call centre were recorded one Wednesday morning.
The table shows the lengths of the first 100 calls.

Length, x (mins)	Frequency
$0 < x \leqslant 1.0$	12
$1.0 < x \leqslant 2.0$	16
$2.0 < x \leqslant 4.0$	38
$4.0 < x \leqslant 6.0$	22
$6.0 < x \leqslant 10.0$	12

Complete the histogram to show this information.

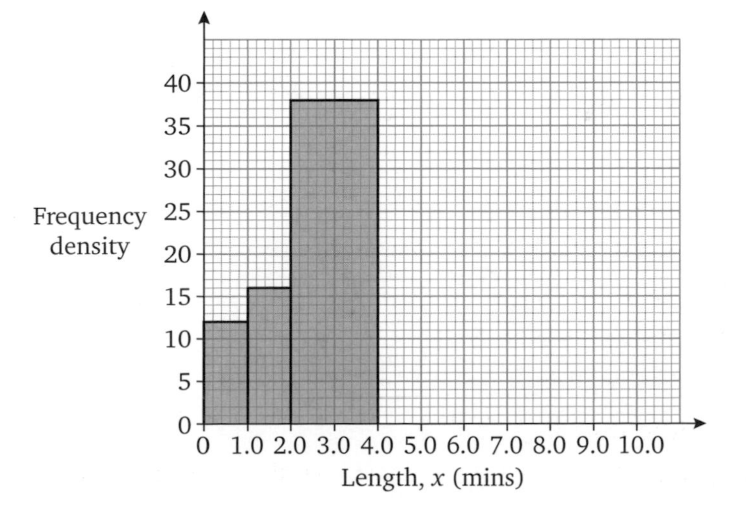

(2 marks)

A
A*

b This histogram represents the lengths of 100 calls on Wednesday evening.

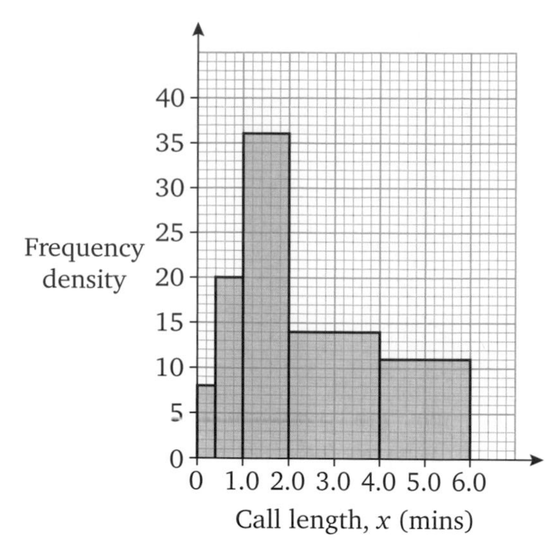

Jake says 'There were more calls lasting less than one minute in the evening than in the morning.'

Is he correct?

You **must** show your working. *(2 marks)*

AQA 2009

A*

13 The table below shows details about passengers on a bus.

	Under 65	**65 or over**
Male	12	5
Female	18	5

The bus company wants to survey two people from the bus.

They choose two at random. Calculate the probability that

a both people chosen are under 65 *(3 marks)*

b one person is male and one is female. *(3 marks)*

Algebra

1 Sequences and symbols

Unit 2

2 Sequences
5 Working with symbols

Unit 3

3 Working with symbols

Key terms

Write down definitions for the following words. Check your answers in the glossary of your Student Book.

expand
factorise
linear sequence
*n*th term
quadratic expression
sequence
term-to-term

Revise... Key points

Sequences Unit 2

The terms of a **sequence** follow a rule.

The **term-to-term** rule for the sequence 3, 7, 11, 15, 19, ... is +4 because you add 4 to the last term to get the next one.

This is an ascending sequence because the numbers are going up.

The **term-to-term** rule for the sequence 15, 8, 1, −6, −13, ... is −7.

This is a descending sequence because the numbers are going down.

Both sequences are called **linear sequences** because the differences are always the same.

To find the ***n*th term** of a linear sequence, first find the differences.

Then use the formula:

$$n\text{th term} = \text{difference} \times n + (\text{first term} - \text{difference})$$
$$= dn + (a - d) \text{ using } d \text{ for the difference and } a \text{ for the first term}$$

For the sequence 15, 8, 1, −6, −13, ... the *n*th term will be $-7n + (15 - -7) = 22 - 7n$

Check that substituting $n = 2$ gives you the correct second term:
$22 - 7 \times 2 = 8$ ✓

Using brackets Units 2 3

When you **expand** (or multiply out) brackets, you must multiply all the terms inside the brackets by the term outside the brackets.

For example, when you expand $3a(3 - 2a)$ the $3a$ multiplies both the 3 and the $-2a$.

Solution: $3a(3 - 2a) = 9a - 6a^2$ Remember that $3a \times -2a = -6a^2$

AQA Examiner's tip

After factorising, always check your answer by multiplying out the brackets.

Factorising is the opposite of **expanding**.

To factorise $6y^2 + 9y$, look for the common factor of $6y^2$ and $9y$, which is $3y$.

Solution: $6y^2 + 9y = 3y(2y + 3)$

When you multiply **two brackets together**, each term in the first bracket multiplies each term in the second bracket. The multiplication can be worked out in a table as shown in this example.

Multiply out and simplify $(x - 5)(x - 2)$

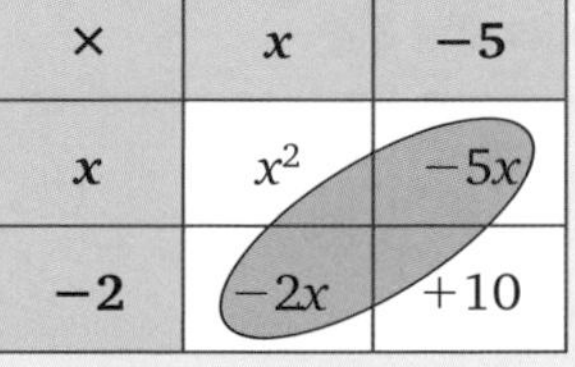

×	x	-5
x	x^2	$-5x$
-2	$-2x$	$+10$

$(x - 5)(x - 2) = x^2 - 5x - 2x + 10 = x^2 - 7x + 10$

The result is a **quadratic expression:** $x^2 - 7x + 10$

There are several ways of setting out this multiplication – use whichever you find works well for you.

Using algebraic fractions Unit 2

Algebraic fractions obey the same rules as ordinary fractions.

Addition and subtraction uses common denominators: $\frac{x}{2} + \frac{x}{3} = \frac{3x}{6} + \frac{2x}{6} = \frac{5x}{6}$

For multiplication, multiply the numerators, multiply the denominators, then cancel any common factors:

$$\frac{2x}{3} \times \frac{3y}{4} = \frac{6xy}{12} = \frac{xy}{2}$$

To simplify an algebraic fraction, first factorise the numerator and/or the denominator.

Then divide by any common factor:

$$\frac{3x - 9}{2xy - 6y} = \frac{3\cancel{(x - 3)}}{2y\cancel{(x - 3)}} = \frac{3}{2y}$$

Here the common factor is the bracket $(x - 3)$

AQA Examiner's tip

If there are no common factors, go back and check your working. There should always be a common factor when you have been asked to simplify a fraction.

Example Sequences Unit 2

C

The first four terms of a sequence are −8, −3, 2, 7.

Find the nth term.

Bump up your grade

To get a Grade C, you need to be able to find the nth term of a sequence.

Solution

−8, −3, 2, 7

+5 +5 +5

The difference is +5 and the first term is −8.

Using the formula $dn + (a - d)$, the nth term is $5n + (-8 - 5) = 5n - 13$

Alternative method

The term-to-term rule is +5
so the nth term is $5n + \ldots$

First term: $5 \times 1 + \ldots = -8$ Think: $5 + ? = -8$

Second term: $5 \times 2 + \ldots = -3$ The missing number is −13

Third term: $5 \times 3 + \ldots = 2$ Check this in the second term: $10 + -13 = -3$ ✓

The nth term is $5n - 13$

Example Symbols Units 2 3

C

1 Expand and simplify $4(2x - y) - 3(x - 2y)$

Solution

$$4(2x - y) - 3(x - 2y) = 8x - 4y - 3x + 6y$$
$$= 5x + 2y$$

When you have a negative term (like −3) in front of the bracket, remember to change the sign of the second term.

Example Symbols (Units 2 3)

B

2 Expand and simplify $(2x + 5)(x - 1)$

Solution

This solution uses the FOIL method: First, Outer, Inner, Last.

$(2x + 5)(x - 1)$ (F, O) $F = 2x \times x = 2x^2$

$O = 2x \times -1 = -2x$

$(2x + 5)(x - 1)$ (I, L) $I = +5 \times x = +5x$

$L = +5 \times -1 = -5$

$$(2x + 5)(x - 1) = \overset{F}{2x^2} - \overset{O}{2x} + \overset{I}{5x} - \overset{L}{5}$$

$$= 2x^2 + 3x - 5$$

Example Symbols (Unit 2)

B

3 Simplify $\frac{3x}{4} + \frac{x}{5} - \frac{5x}{8}$

Solution

The denominators are 4, 5 and 8.

The lowest common denominator is 40. because it is a multiple of 4, 5 and 8

$\frac{3x}{4} = \frac{30x}{40}$ To change $\frac{3x}{4}$ to $\frac{?}{40}$ multiply top and bottom by 10.

$\frac{x}{5} = \frac{8x}{40}$ To change $\frac{x}{5}$ to $\frac{?}{40}$ multiply top and bottom by 8.

$\frac{5x}{8} = \frac{25x}{40}$ To change $\frac{5x}{8}$ to $\frac{?}{40}$ multiply top and bottom by 5.

$$\frac{3x}{4} + \frac{x}{5} - \frac{5x}{8} = \frac{30x}{40} + \frac{8x}{40} - \frac{25x}{40} = \frac{13x}{40}$$

Practise... Sequences and symbols (Units 2 3)

D C B

D

1 The nth term of a sequence is $5n - 1$

a Write down the first three terms of the sequence.

b Show that the 17th term is a multiple of the 3rd term.

c Explain why 205 cannot be a number in this sequence.

2 Expand **a** $4(2k - 1)$ **b** $m(6 - m)$ **c** $2t(3t + 2)$

3 Factorise **a** $10x + 5y$ **b** $t^2 - 5t$ **c** $3k^2 - 9k$

4 The area of this rectangle is $(8p + 4)$ cm².

4 cm

Not drawn accurately.

What is the length of the rectangle?

C

5 Write down the nth term for the following linear sequences.

a 2, 5, 8, 11, …

b $1, 3\frac{1}{2}, 6, 8\frac{1}{2}, \ldots$

6 Expand and simplify:

a $3(2a + 3b) + 4(a - b)$ **b** $2(x - 3y) - 3(2y - x)$

7 Here is a sequence of numbers: 3, 4, 7, 16, …

The rule for continuing this sequence is: multiply by 3 and subtract 5.

What are the next two numbers in this sequence?

8 x and y are integers.

x is odd and y is even.

Decide whether each statement is true or false.

i $x + y$ is always odd.

ii $2x - y$ is always odd.

iii $5xy$ is always even.

iv $2x + 4y$ is always a multiple of 4.

B

9 Expand and simplify:

a $(x + 3)(x + 8)$ **d** $(2x + 1)(x - 5)$ **g** $(x + 7)^2$

b $(y - 2)(y - 1)$ **e** $(3y - 5)(y + 2)$ **h** $(2y - 1)(2y + 1)$

c $(z - 4)(z + 3)$ **f** $(z + 6)(4z - 5)$ **i** $(3t - 2)^2$

AQA Examiner's tip

When you have to square a bracket, write it out in full. For example $(x + 7)^2 = (x + 7)(x + 7)$

10 Simplify these fractions.

a $\frac{a}{3} + \frac{a}{5}$ **c** $\frac{4cd}{5} \times \frac{e}{2c}$ **e** $\frac{6m + 4n}{9m + 6n}$

b $\frac{7b}{8} - \frac{5b}{12}$ **d** $\frac{pq}{3t} \times \frac{pt}{q}$ **f** $\frac{12}{4p - 4q}$

Algebra

2 Equations and formulae

Unit 2

7 Equations and inequalities
11 Formulae

Unit 3

6 Equations and formulae

Key terms

Write down definitions for the following words. Check your answers in the glossary of your Student Book.

brackets
denominator
equation
expand
expression
formula
identity
integer
solution
solve
subject
substitute
substitution
unknown

Revise... Key points

Equations

Units 2 3

You will be asked to **solve** an equation to find the **unknown.**

For example, the equation $3y + 7 = 13$ can be solved to find the **solution** $y = 2$. Here the unknown is y.

When the unknown appears on both sides of the equation, collect together on one side all the terms that contain the unknown. Collect the terms that do **not** contain the unknown on the other side.

For example: $5x + 2 = 7x - 3$

$7x$ is greater than $5x$ so collect all the terms in x on the right-hand side.

$5x + 2 - 5x = 7x - 3 - 5x$ taking $5x$ from both sides

$2 = 2x - 3$

$2 + 3 = 2x - 3 + 3$ adding 3 to both sides

$5 = 2x$

$2.5 = x$ dividing both sides by 2

$x = 2.5$ Always give your answer as $x = \dots$

AQA Examiner's tip

If one of the terms in x has a negative coefficient, remember that, for example, $-3x$ is smaller than $(+)x$.

For the equation $7 + x = 4 - 3x$, collect terms in x on the **left**.

When the equation contains **brackets**, start by dealing with the brackets.

This usually means multiplying out (or **expanding**) the brackets.

For example: $4(d - 3) = 5$

$4d - 12 = 5$ after multiplying out the brackets

$4d = 17$ after adding 12 to both sides

$d = 4.25$ after dividing by 4

When the equation contains fractions, clear the fractions by multiplying by the lowest common **denominator**.

Reminder: the lowest common denominator is the smallest number that is a multiple of both denominators.

For example: $\frac{2m}{3} - 7 = \frac{m}{6}$

Lowest common denominator is 6.

Multiply every term by 6.

$\overset{2}{\cancel{6}} \times \frac{2m}{\cancel{3}} - 6 \times 7 = \cancel{6} \times \frac{m}{\cancel{6}}$ Don't forget to multiply the term -7 by 6 as well as the fraction terms.

$4m - 42 = m$

$3m = 42$

$m = 14$

Formulae

Units 2 3

A **formula** will usually be written in symbols such as $V = l \times w \times h$ for the volume of a cuboid.

V is on the left-hand side and is called the **subject** of the formula.

This formula tells you how to work out the volume when you know the values of other quantities (length, width and height). Putting these values into the formula is called **substitution**.

Using the formula above, if $l = 10$, $w = 5$ and $h = 3$ then $V = 10 \times 5 \times 3 = 150$

In Unit 2, you may be asked to change the subject of a formula: for example, if you want to find the length of a cuboid and you know the width and height. The formula $V = l \times w \times h$ has to be changed round so that l is the subject.

To change the subject of a formula, use the same steps as in solving an equation.

$V = l \times w \times h$ Divide both sides by w and h to leave l on its own.

$\frac{V}{wh} = l$ Now swap the sides to get l on the left.

$l = \frac{V}{wh}$

Distinguishing formulae, expressions, equations and identities

Units 2 3

Like an **equation**, a formula always includes an equals sign. A formula is true for a range of values, whereas equations are true only for certain values, which you find by solving the equation.

An **expression** does not contain an equals sign.

An **identity** is always true whatever the value of the symbols.

- $5x + 2$ is an expression.
- $5x + 2 = 17$ is an equation: it is only true when $x = 3$
- $P = 5x + 2$ is a formula for finding P when you know the value of x
- $5y + 3 \equiv 3 + 5y$ is an identity. $=$ is replaced by $\equiv$ to show it is always true.

Example Equations

Unit 2

C

Solve the equation $3(2x - 5) = 13 - 2(5 - x)$

Bump up your grade

To get a Grade C, you need to be confident solving equations with brackets.

Solution

Start by multiplying out the brackets.

$3(2x - 5) = 13 - 2(5 - x)$ Remember that -2 in front of the second bracket multiplies $-x$ to become $+2x$.

$6x - 15 = 13 - 10 + 2x$

$6x - 15 = 3 + 2x$

Collect terms in x on the left and terms without x on the right.

$6x - 15 - 2x = 3 + 2x - 2x$

$4x - 15 = 3$

$4x - 15 + 15 = 3 + 15$

$4x = 18$

$x = 4\frac{1}{2}$

Example Equations Unit 3

C

Triangle PQR is isosceles. One angle is $x°$.
Another angle is $7x°$.
Find the two possible values of x.

Sometimes you will be required to form the equation and then solve it, as in this example.

Solution

In an isosceles triangle, two angles are equal.

The angles could be $x°$, $x°$ and $7x°$ or they could be $7x°$, $7x°$ and $x°$.

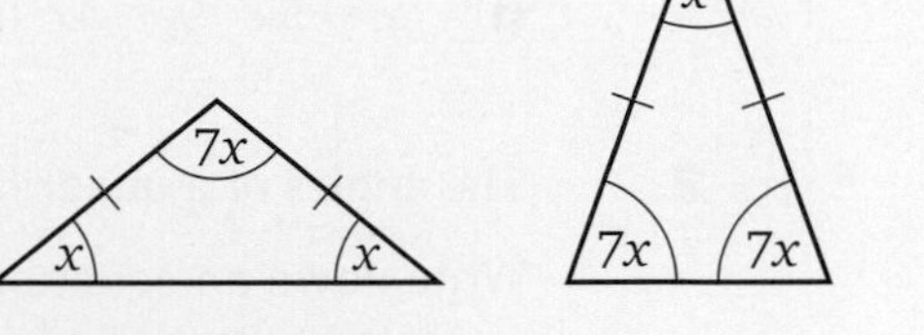

$x + x + 7x = 180$ $\quad\quad$ $7x + 7x + x = 180$

$9x = 180$ $\quad\quad$ $15x = 180$

$x = 20$ $\quad\quad$ $x = 12$

Check the first answer: $20 + 20 + 140 = 180$ ✓

Check the second answer: $84 + 84 + 12 = 180$ ✓

AQA Examiner's tip

Drawing a diagram often helps you to solve a problem.
Take the time to draw it neatly and legibly – it might earn you method marks.

Example Formulae Unit 3

C

The formula for finding the area of a ring is $A = \pi(R^2 - r^2)$ where R is the radius of the outer circle and r is the radius of the inner circle. Find the area when $R = 8.5$ cm and $r = 7$ cm.
Give your answer correct to the nearest square centimetre.

Solution

$A = \pi(R^2 - r^2)$ $\quad$ Copy the formula.

$A = \pi(8.5^2 - 7^2)$ $\quad$ Substitute (replace each letter with its value).

$= \pi \times 23.25$

$= 73.04\ldots$

The area is 73 cm².

Example Equations Unit 2

B

Solve the equation $\dfrac{3x + 5}{4} - \dfrac{x - 2}{3} = 4$

Solution

First rewrite the equation, putting in the 'invisible brackets'.

$$\frac{(3x + 5)}{4} - \frac{(x - 2)}{3} = 4$$

The lowest common denominator of 4 and 3 is 12, so multiply by 12.

$$^{3}\cancel{12} \times \frac{(3x + 5)}{\cancel{4}} - {^{4}\cancel{12}} \times \frac{(x - 2)}{\cancel{3}} = 12 \times 4$$ Make sure that you multiply the right-hand side by 12.

$3(3x + 5) - 4(x - 2) = 48$ $\quad$ After cancelling – the denominators should no longer appear.

$9x + 15 - 4x + 8 = 48$ $\quad$ Remember -4 multiplies -2 to give $+8$.

$5x + 23 = 48$

$5x = 25$

$x = 5$

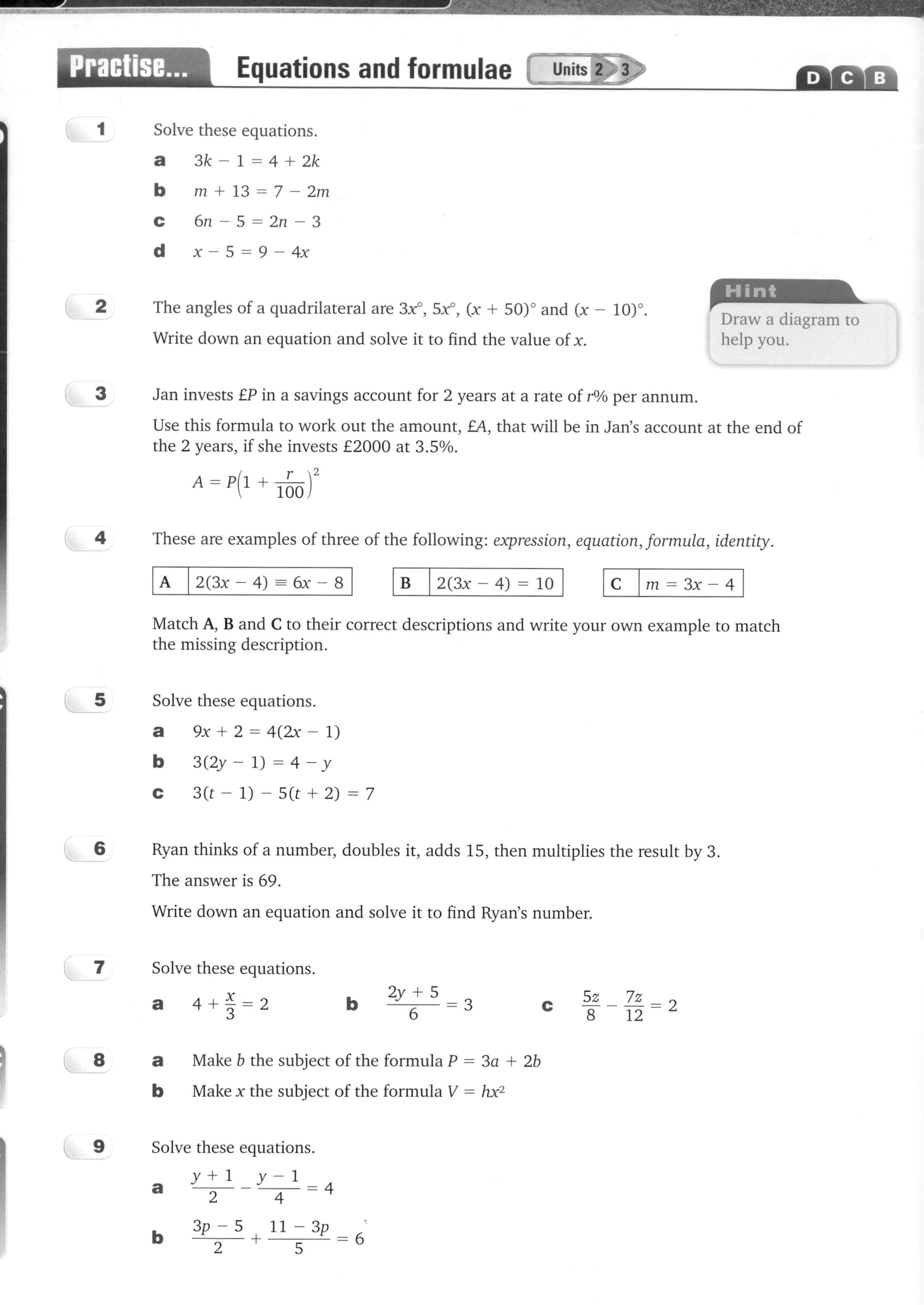

Practise... Equations and formulae

Units 2 3

D C B

D

1 Solve these equations.

a $3k - 1 = 4 + 2k$

b $m + 13 = 7 - 2m$

c $6n - 5 = 2n - 3$

d $x - 5 = 9 - 4x$

2 The angles of a quadrilateral are $3x°$, $5x°$, $(x + 50)°$ and $(x - 10)°$.

Write down an equation and solve it to find the value of x.

Hint

Draw a diagram to help you.

3 Jan invests £P in a savings account for 2 years at a rate of r% per annum.

Use this formula to work out the amount, £A, that will be in Jan's account at the end of the 2 years, if she invests £2000 at 3.5%.

$$A = P\left(1 + \frac{r}{100}\right)^2$$

4 These are examples of three of the following: *expression, equation, formula, identity*.

A	$2(3x - 4) \equiv 6x - 8$

B	$2(3x - 4) = 10$

C	$m = 3x - 4$

Match **A**, **B** and **C** to their correct descriptions and write your own example to match the missing description.

C

5 Solve these equations.

a $9x + 2 = 4(2x - 1)$

b $3(2y - 1) = 4 - y$

c $3(t - 1) - 5(t + 2) = 7$

6 Ryan thinks of a number, doubles it, adds 15, then multiplies the result by 3.

The answer is 69.

Write down an equation and solve it to find Ryan's number.

7 Solve these equations.

a $4 + \frac{x}{3} = 2$ **b** $\frac{2y + 5}{6} = 3$ **c** $\frac{5z}{8} - \frac{7z}{12} = 2$

C B

8 **a** Make b the subject of the formula $P = 3a + 2b$

b Make x the subject of the formula $V = hx^2$

B

9 Solve these equations.

a $\frac{y + 1}{2} - \frac{y - 1}{4} = 4$

b $\frac{3p - 5}{2} + \frac{11 - 3p}{5} = 6$

Algebra

3 Trial and improvement

Unit 3
13 Trial and improvement

Key terms

Write down definitions for the following words. Check your answers in the glossary of your Student Book.

round
solve/solution
trial and improvement

Revise... Key points

Trial and improvement

Trial and improvement is used to **solve** equations by using **estimations** which get closer and closer to the **solution**. On the examination paper you will be told when to use trial and improvement. Do **not** use it, for example, to solve simultaneous equations. You will be asked to give a **rounded** answer – read the question carefully and when you have finished the question, check that your answer has been rounded correctly.

It is best to set out your working in a table as shown in the example.

Bump up your grade

When you can solve the equations in this chapter by trial and improvement, you are working at Grade C.

Example Trial and improvement Unit 3

C

Use trial and improvement to find the solution of the equation $x^3 - 5x = 62$ that lies between 4 and 5.

Give your answer to one decimal place.

Solution

Set up the table, with the three headings shown below.

Start by trialling the values given in the question, $x = 4$ and $x = 5$.

x	$x^3 - 5x$	Comment
4	64 − 20 = 44	too small
5	125 − 25 = 100	too large

These results suggest that the solution is closer to 4 than to 5, because 44 is closer to 62 than 100 is.

For your next trial, you could use $x = 4.3$ or 4.4

Here are two more trials.

4.3	79.507 − 21.5 = 58.007	too small
4.4	85.184 − 22 = 63.184	too large

The number 62 lies between 58.007 and 63.184

So the solution lies between 4.3 and 4.4

To find out whether it is closer to 4.3 or 4.4, trial at $x = 4.35$ halfway between 4.3 and 4.4

4.35	82.312 ... − 21. 75 = 60.562 ...	too small

The number 62 lies between 60.562... and 63.184

So the solution lies between 4.35 and 4.4

To one decimal place, the solution is 4.4

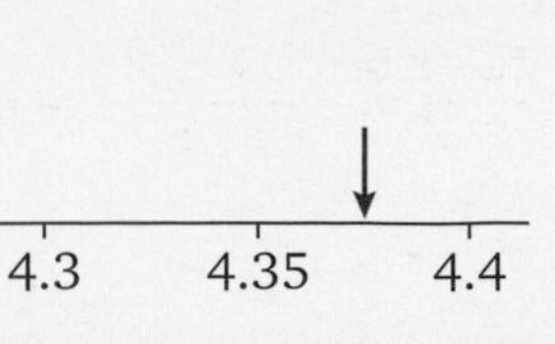

AQA Examiner's tip

Write out your calculations clearly, with at least two decimal places.

There will be marks given for this working so you must make sure that the examiner can read it.

Practise... Trial and improvement Unit 3

C

1 A solution of the equation $x^3 + x = 8$ lies between $x = 1$ and $x = 2$.

Use trial and improvement to find this solution to one decimal place.

2 A solution of the equation $y^3 - 3y = 7$ lies between $y = 2$ and $y = 3$.

Use trial and improvement to find this solution to one decimal place.

3 A solution of the equation $t^3 - t^2 = 52$ lies between $t = 4$ and $t = 5$.

Use trial and improvement to find this solution to two decimal places.

4 A solution of the equation $m^3 - 7m = 5$ lies between $m = -3$ and $m = -2$.

a Use trial and improvement to find this solution to one decimal place.

b Use trial and improvement to find a positive solution of this equation. Give your answer correct to one decimal place.

5 A solution of the equation $x^2 + \frac{2}{x} = 19$ lies between $x = -5$ and $x = -4$.

Use trial and improvement to find this solution to one decimal place.

6 A solution of the equation $p^3 - 3p^2 = 175$ lies between $p = 6$ and $p = 7$.

Use trial and improvement to find this solution to one decimal place.

Hint

Remember that $3p^2 = 3 \times p \times p$, not $3p \times 3p$.

7 The dimensions of this cuboid are in centimetres.

The volume of the cuboid is 120 cm³.

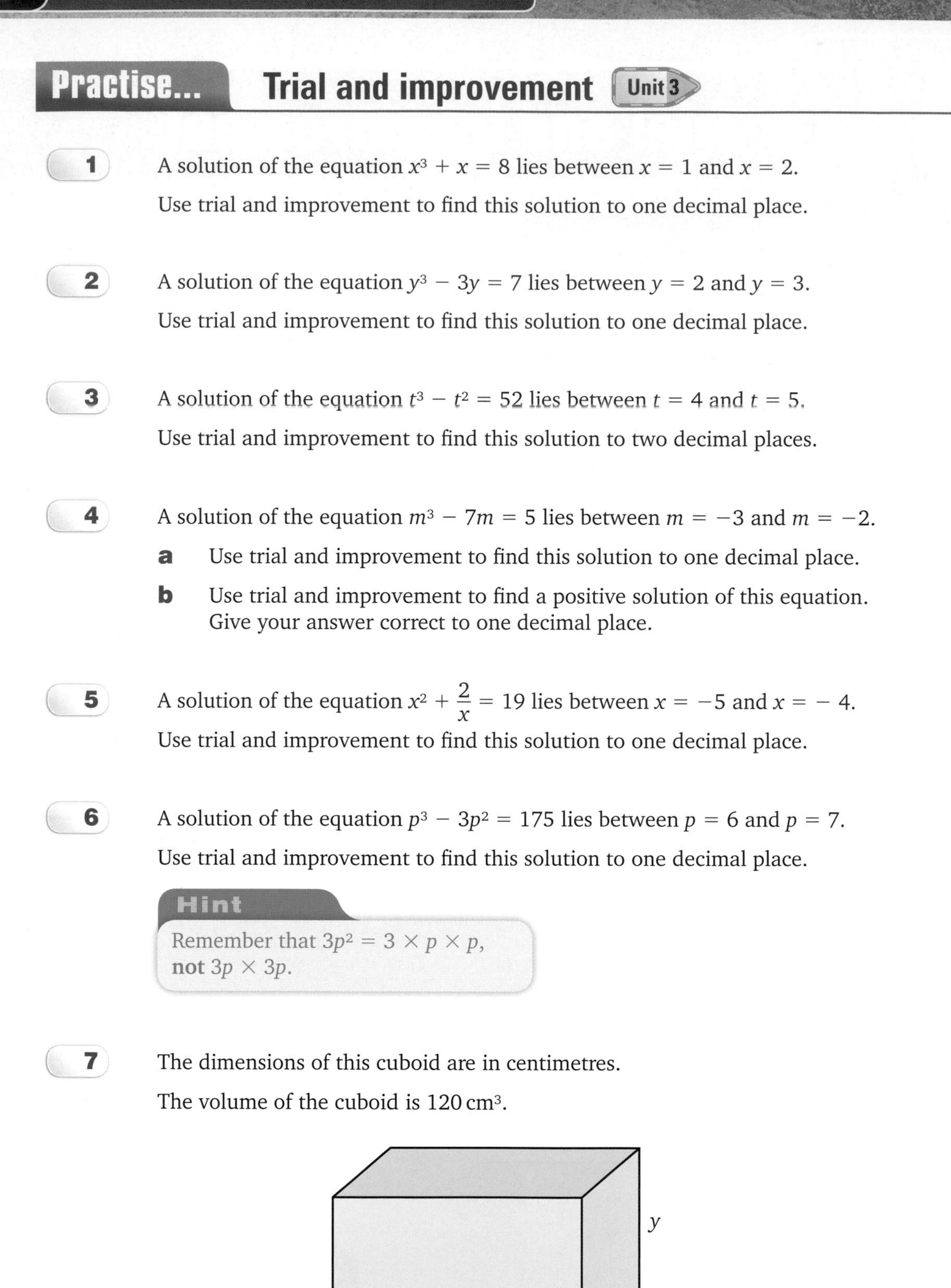

Find the value of y, correct to one decimal place.

Algebra

4 Coordinates and graphs

Unit 2

6 Graphs of linear functions
10 Real-life graphs

Unit 3

8 3-D shapes, coordinates and graphs

Key terms

Write down definitions for the following words. Check your answers in the glossary of your Student Book.

coefficient
constant
gradient
intercept
linear
speed
variable

Revise... Key points

Drawing straight-line graphs Unit 2

An equation such as $y = 4 - 5x$ is called a **linear** equation. It does not contain any powers of x or y such as x^2 or y^3.

When you draw the graph of an equation like this, it will always be a straight line.

To draw the graph, plot three points whose coordinates fit the equation of the line.

You first choose three values of x, then use the equation to find the values of y.

$x = 0$ is a good choice because it is easy to substitute in the equation.

If you choose the endpoints of the range you are given as your other two values of x, you will know what range is needed on the y-axis.

AQA Examiner's tip

If your three plots are **not** in a straight line, go back and check your working for finding the value of y. You have made a mistake!

Finding the gradient and intercept Unit 2

The **gradient** of a straight-line graph is a measure of how steep it is.

You can find the gradient either from the graph itself or from its equation.

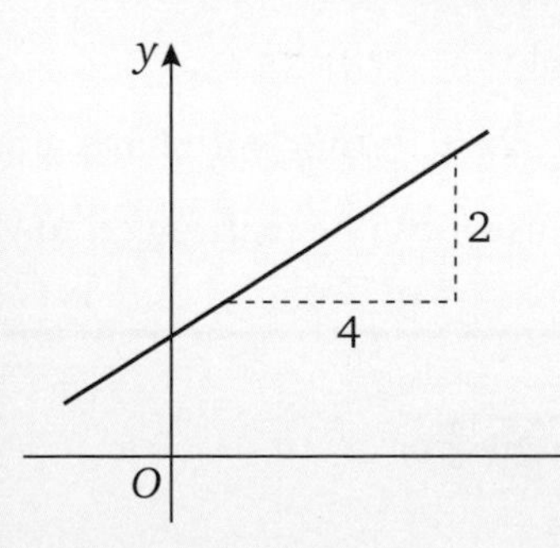

The gradient of this line is $\frac{2}{4} = \frac{1}{2}$

You can draw the triangle anywhere along the length of the graph.

Make the base of the triangle a whole number so it is easy to do the division.

The gradient is positive because as x **increases**, y also **increases**.

AQA Examiner's tip

If you are working from the graph, check the scales used on the graph, as they may not be the same on both axes.

To find the gradient from the equation, you need to have the equation in the form $y = mx + c$.

y and x are the **variables** in the equation.

The **coefficient** of x (m) is the gradient of the line.

c is the **constant** in the equation of the line.

So the gradient of $y = 4x - 7$ is 4 and the constant is -7.

When $x = 0$, $y = c$ so the line crosses the y-axis at $(0, c)$.

This point is called the **intercept**.

If $c = 0$, the line goes through the origin.

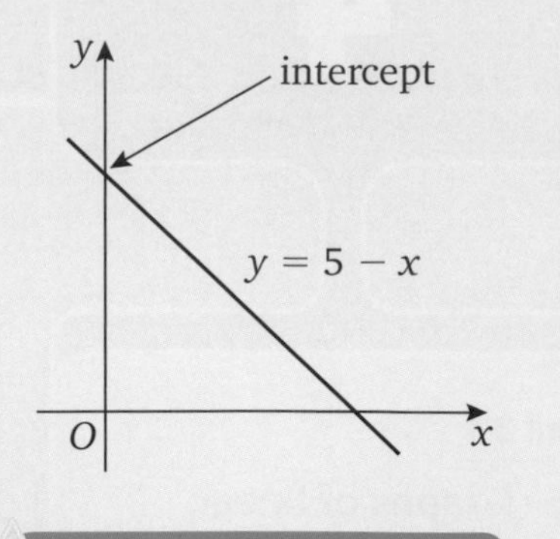

The gradient of $y = 5 - x$ is -1
(You can rewrite the equation as $y = -x + 5$)

A line with a negative gradient slopes from top left to bottom right; as x **increases**, y **decreases**.

Bump up your grade

When you can find a gradient, you are working at Grade C.

Finding the midpoint of a line segment Unit 2

A **line segment** is defined by its endpoints, for example: from $P(1, 4)$ to $Q(5, -2)$.

The midpoint of PQ can be found from an accurate drawing on graph paper.

It can also be found by calculating the mean of the coordinates of P and Q.

The midpoint is $\left(\frac{1+5}{2}, \frac{4+-2}{2}\right) = (3, 1)$.

Lines through two given points

To find the equation of a line through two given points, first find the gradient.

Once you know the gradient, use the coordinates of one of the points to work out the intercept.

Parallel lines Unit 2

Lines that have the same gradient are parallel.

The line $y = 2x + 5$ is parallel to the line $y = 2x - 7$; they both have gradient 2.

The line $y - 2x = 1$ also has gradient 2 because it can be rearranged to $y = 2x + 1$

Distance–time graphs Units 2 3

Distance–time graphs tell you about a journey.

Time is measured along the horizontal axis and distance is measured along the vertical axis.

Time may be measured from the start of the journey or be given as actual time using am and pm or the 24-hour clock.

The distance is measured from a particular point, usually the starting point.

If the graph goes back to the horizontal axis, the traveller is returning to the starting point.

The gradient of the line on a distance–time graph gives the speed of travel.

If the graph is horizontal (no increase in the distance or direction), the traveller has stopped.

Other real-life graphs (Units 2 3)

There are many uses for graphs in real-life situations. Here are some examples.

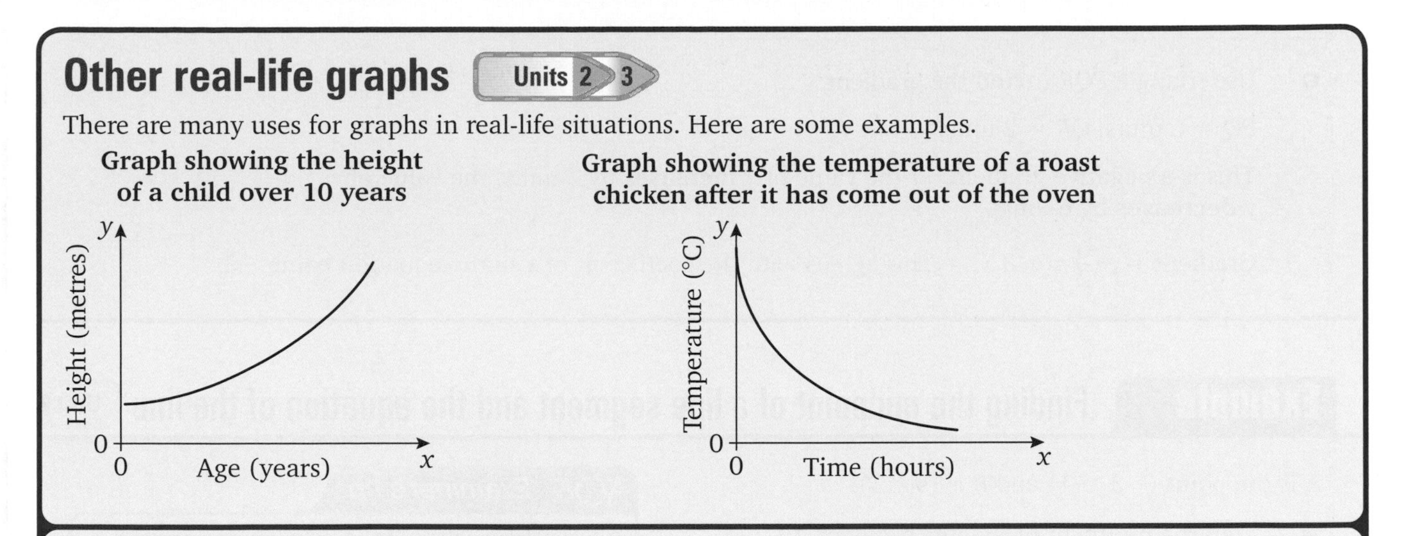

Coordinates in 3-D (Unit 3)

Coordinates in two dimensions give the position of a point in a **plane**. For example on the page of a book or on the floor of a room.

Coordinates in three dimensions give the position of a point in a space. For example a light bulb hanging from the ceiling.

The third coordinate is the z-coordinate, which tells you how high the point is above the base.

In this diagram of a cuboid, O is the origin.

$OA = 2$ units, $AB = 6$ units and $AD = 5$ units.

A is on the x-axis, at $x = 2$. Its coordinates are (2, 0, 0)

x-coordinate
y-coordinate
z-coordinate

B has the same x-coordinate as A and its y-coordinate is 6.

The coordinates of B are (2, 6, 0).

C is 5 units vertically above B. The coordinates of C are (2, 6, 5).

D is 5 units vertically above A. The coordinates of D are (2, 0, 5).

Example: Drawing straight-line graphs (Units 2 3)

DC

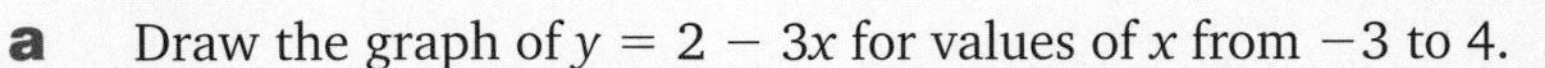

a Draw the graph of $y = 2 - 3x$ for values of x from -3 to 4.

b Write down the coordinates of the point where this line crosses the line $y = -6$.

c Use your graph to find the gradient of the line.

Solution

a Choose 3 values for x:

$x = 0$, $x = -3$ and $x = 4$

x	-3	0	4
y	11	2	-10

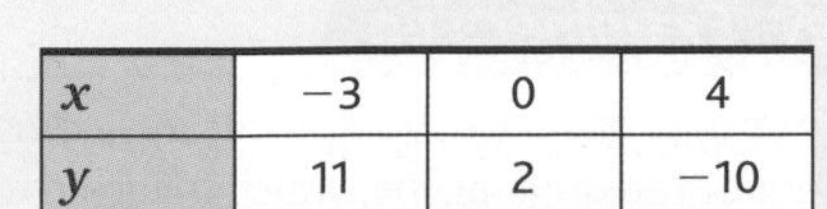
AQA Examiner's tip

$x = 0$ is a good choice because it is easy to substitute in the equation.

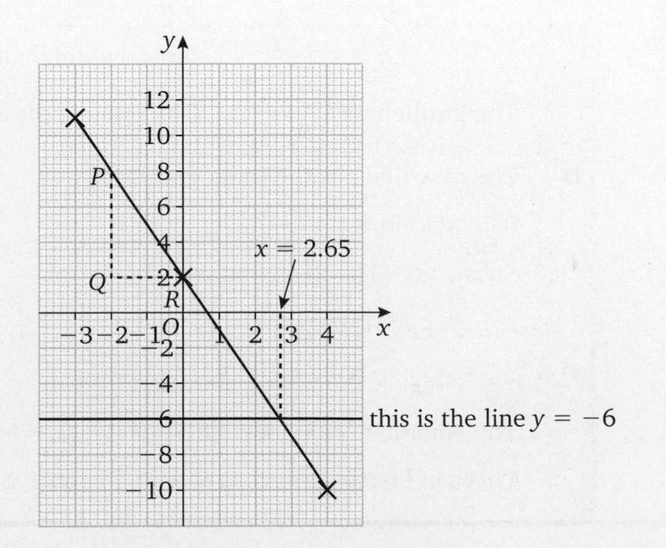

b The graph crosses $y = -6$ at $(2.65, -6)$.

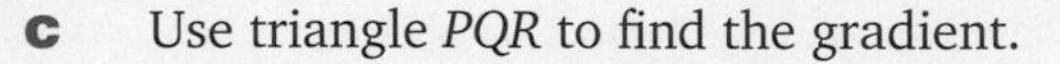

c Use triangle PQR to find the gradient.

$PQ = 6$ units, $QR = 2$ units

This is a negative gradient: as the value of x **increases** by 2 units, the value of y **decreases** by 6 units.

Gradient $= -\frac{6}{2} = -3$ This agrees with the coefficient of x in the equation being -3.

Example Finding the midpoint of a line segment and the equation of the line Unit 2

C B

A is the point $(-3, -1)$ and B is $(9, 11)$.

a Find the midpoint of the line segment AB.

b Find the gradient of the line AB.

c Find the equation of the line AB.

AQA Examiner's tip

Always draw a diagram to help you to answer a question of this type. Draw it clearly so it will help you to check whether your answer is reasonable. It may also help you to gain method marks.

Solution

Sketch the positions of A and B on a grid.

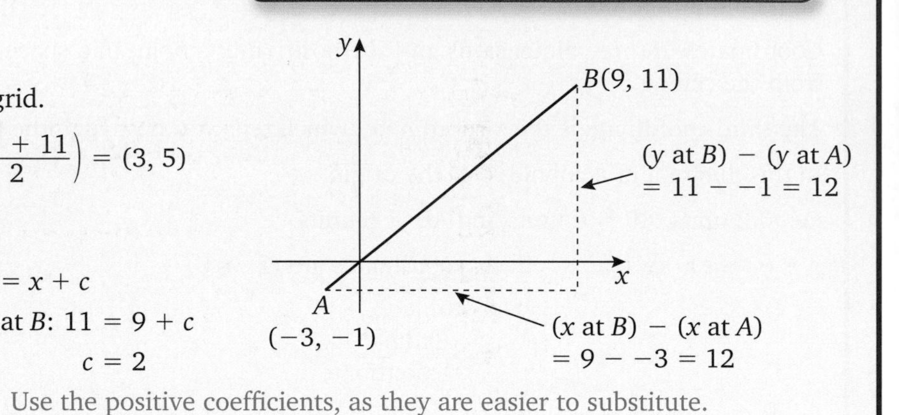

a The midpoint is at $\left(\frac{-3+9}{2}, \frac{-1+11}{2}\right) = (3, 5)$

b The gradient of AB is $\frac{12}{12} = 1$

c The equation of the line AB is $y = x + c$

Substitute the values of x and y at B: $11 = 9 + c$

$c = 2$

AB is the line $y = x + 2$ Use the positive coefficients, as they are easier to substitute.

Example Parallel lines Unit 2

D C

a Find the gradient and the intercept of the line $x + 4y = 5$

b Find the equation of the line through $(1, -2)$ that is parallel to $x + 4y = 5$

Solution

a To find the gradient of $x + 4y = 5$, make y the subject of the equation.

$x + 4y = 5$

$4y = 5 - x$ after taking x from both sides

$y = \frac{5}{4} - \frac{x}{4}$ after dividing both sides by 4

The gradient is $-\frac{1}{4}$ The intercept is at $\left(0, \frac{5}{4}\right)$

b The new line has the same gradient.

Its equation is $y = -\frac{x}{4} + c$

Substitute $(1, -2)$ to find c.

$-2 = -\frac{1}{4} + c$

$c = -2 + \frac{1}{4} = -\frac{7}{4}$

The equation of the new line is $y = -\frac{x}{4} + -\frac{7}{4}$

This can be rearranged as $4y = -7 - x$ or $x + 4y = -7$

AQA Examiner's tip

This example has been worked out by finding the gradient of the original line, which involved changing the subject. Because the new line has the same gradient, it has the same terms in x and y so it starts $x + 4y = \ldots$

Fill in the last term by substituting $(1, -2)$ in $x + 4y$

This is a useful 'shortcut'.

Example Distance–time graph Unit 2

This graph shows Ginny's journey to work.
She leaves home at 0700 and cycles to the station.
She catches a train to the town where she works.
When she gets off the train, she walks to her office.

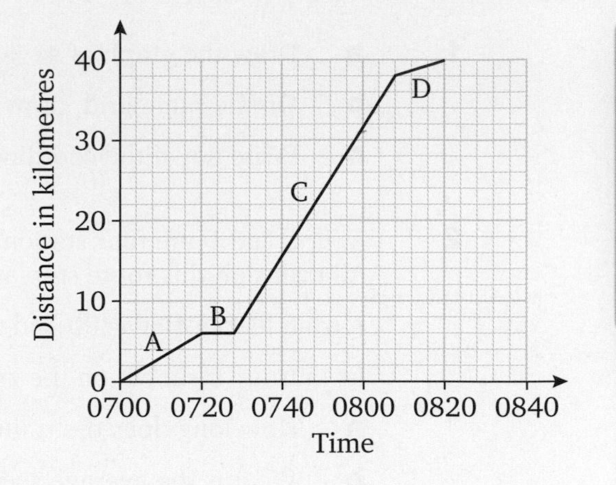

a Describe what is happening at section *B* of the graph.

b How far does Ginny travel on the train?

c Work out the average speed of the train in km/h.

D C

Solution

a Section *B* of the graph represents Ginny waiting for the train.
On the horizontal scale, each small square represents 4 minutes, so Ginny waits for 8 minutes.

b Section *C* of the graph represents Ginny travelling on the train.
On the vertical scale, each small square represents 2 km.
Her train journey starts 6 km from home and finishes 38 km from home, so she travels 32 km on the train.

c The train travels 32 km between 0728 and 0808, which is 40 minutes.

$$\text{Speed} = \frac{\text{Distance}}{\text{Time}} = \frac{32}{40/60}\text{ km/h} = 48\text{ km/ha}$$

AQA Examiner's tip

It is often easier to work out speeds using proportions.
In 40 min, the train travels 32 km.
In 20 min, it travels 16 km.
In one hour, it travels $3 \times 16 = 48$ km.
So the speed is 48 km/h.

Example Other real-life graphs Units 2 3

The graph shows how the value of a new car decreases over a number of years.

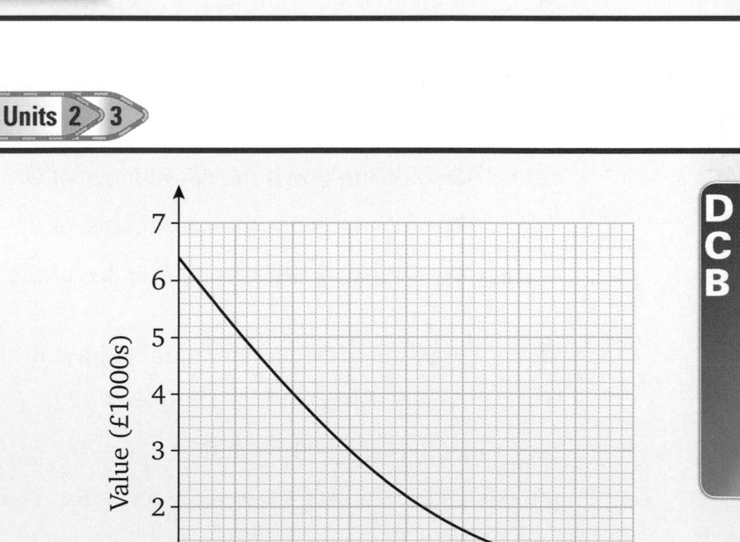

a What did the car cost when new?

b How much is it worth after 4 years?

c The owner sells when the car has lost half its value. When is this?

d Explain why it is not sensible to continue this graph to show the value after 20 years.

D C B

Solution

a Read off the value when the time is zero:
£6400 1 square = £200 on the vertical axis

b Read off the value when the time is 4: £2200

c Read off the time when the value is £3200:
2.7 years = 2 years 8 months 1 square = 0.2 years on the horizontal axis

d The car is unlikely to be on the road after 20 years.
If it is, it probably has only scrap value.

Practise... Coordinates and graphs

Units 2 3

D C B

D

1 **a** Draw the graph of $4x + 3y = 24$ for values of x from 0 to 6.

b On the same grid, draw the graph of $y = \frac{1}{2}x$.

c Write down the coordinates of the point where the two graphs cross.

D C

2 A, B, C and D are four stations on a railway line.
All trains on this route stop at stations B and C.

A train leaves A at 1400 and travels to D.

Its journey is shown on the graph.

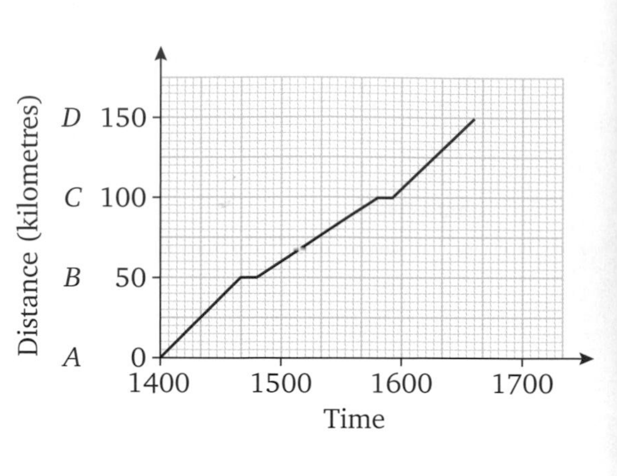

a How long does the train stop at each station?

b What is the average speed of the train between A and B?

c How can you tell by looking at the graph, that the speed is slower from B to C?

A second train leaves D at 1420, going to A.
It stops at C at 1500, then goes on to B at 40 km/h.

d Show this train's journey on the graph.

e When and where do the two trains pass each other?

C

3 M is the midpoint of the line PQ.

P is the point (0, 3).

M is the point (4, 0).

Find the coordinates of Q.

C B

4 A cube of side 6 units is placed with one corner at $P(6, 1, 0)$ as shown in the diagram.

PQ is parallel to the y-axis.

PS is parallel to the x-axis.

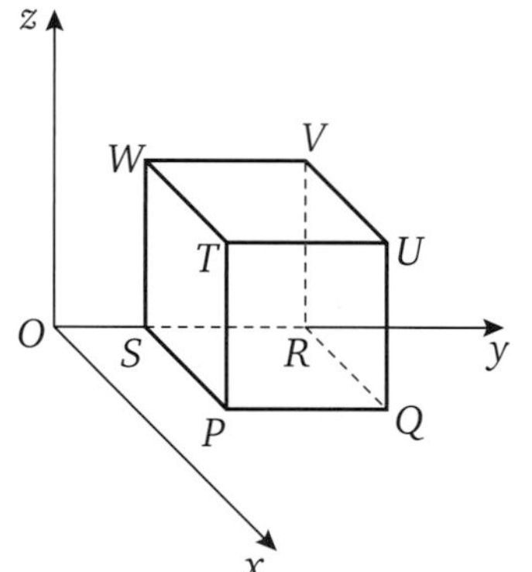

a Write down the coordinates of Q.

b Write down the coordinates of R.

c Write down the coordinates of the midpoint of WU.

5 Find the three pairs of parallel lines in this list.

A: $y = 7 - x$ D: $5x = 3 - 2y$ G: $5y = 3 - x$

B: $2y + 5x = 7$ E: $x - y = 5$ H: $y + 5x = 3$

C: $5y - x = 3$ F: $x + y = 5$ I: $3y + 2 = 3x$

6 **a** Find the gradient of the line joining the points (3, 5) and (5, −1).

b Find the equation of this line.

B

7 Find the equation of the line through (−1, 5) that is parallel to $y + 4x = 5$

Algebra

5 Quadratic functions

Unit 2

12 Quadratic equations and algebraic proof

Unit 3

17 Quadratic functions

Key terms

Write down definitions for the following words. Check your answers in the glossary of your Student Book.

coefficient
expand
factorise
linear
parabola
quadratic expression

Revise... Key points

Graphs of quadratic functions (Unit 3)

A **quadratic function** has the form $y = ax^2 + bx + c$

These are all quadratic functions:

$y = 3x^2 - 5x + 4$, $y = x^2 + 1$, $y = 5x - 2x^2$

If there is no term in x^2 the function will be **linear**.
Remember, the graph of a linear function is a straight line.

The graph of a quadratic function is a **parabola**.

If the **coefficient** of x^2 is positive, the graph will be U-shaped (or part of a U-shape).

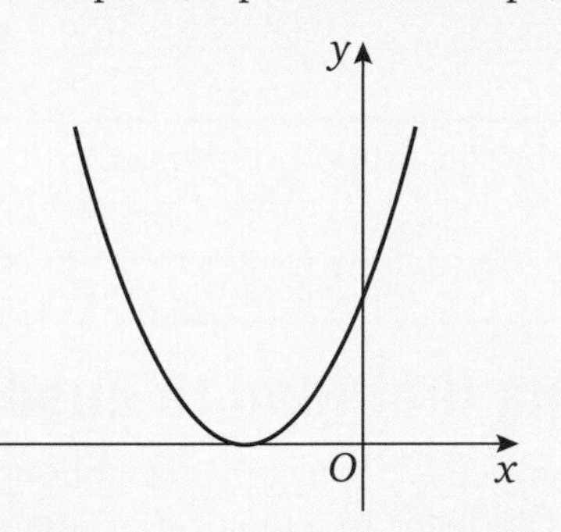

If the coefficient of x^2 is negative, the graph will be hill-shaped.

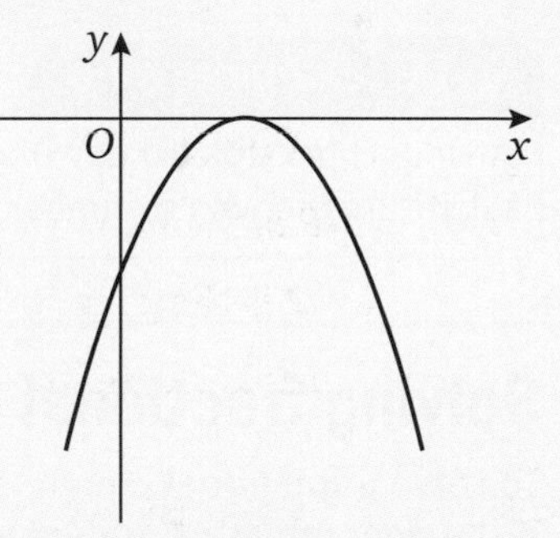

To draw the graph of a quadratic function, you need a table of values.

In the exam, you may be given a table of values to complete, as in the first worked example.

When the values have been plotted, join them with a smooth curve.

AQA *Examiner's tip*

If you join your plotted points with straight lines, you will lose marks. Do not flatten out the bottom (or top) of your curve – keep it curving through the plotted points and make it look symmetrical.

Factorising quadratic expressions (Units 2 3)

When you **expand** brackets like this: $(x - 4)(2x + 3)$ you form a **quadratic expression**: $2x^2 - 5x - 12$

When you are asked to **factorise** a quadratic expression, you reverse this operation to get back to the brackets:

$x^2 - 9x + 20 = (x - 4)(x - 5)$

AQA *Examiner's tip*

The brackets should always contain integers, not fractions or decimals.

Simplifying algebraic fractions Unit 2

First factorise the numerator and the denominator.

Then divide them both by any common factors.

Using the quadratics and their factors introduced above: $\frac{2x^2 - 5x - 12}{x^2 - 9x + 20} = \frac{(x-4)(2x+3)}{(x-4)(x-5)} = \frac{2x+3}{x-5}$

AQA Examiner's tip

If you do not have any common factors, check your working – if you are asked to simplify an algebraic fraction it should always have a common factor.

Solving quadratic equations by factorising Unit 2

This method for solving quadratic equations is based on the fact that if the product of two numbers is zero, one of the numbers must be zero.

To solve $2x^2 - 5x - 12 = 0$, first factorise to get $(x - 4)(2x + 3) = 0$ as shown above

If $(x - 4)(2x + 3) = 0$, then either the first or the second bracket must be equal to zero.

Either $(x - 4) = 0$, which leads to $x = 4$

or $(2x + 3) = 0$, which leads to $2x = -3$, hence $x = -1.5$

The solutions are $x = 4$ and $x = -1.5$

Check by substituting $x = 4$ in the quadratic equation.
$2 \times 4^2 - 5 \times 4 - 12 = 32 - 20 - 12 = 0$ ✓

AQA Examiner's tip

Always check at least one of your answers by substitution. It is easier to substitute the whole number answer here.

Solving fractional equations that lead to quadratics Unit 2

To solve a fractional equation, you usually need to multiply throughout by the lowest common denominator. If the denominators are expressions in x this leads to a quadratic equation.

Solving quadratic equations by completing the square Unit 2

Some quadratics are perfect squares. For example: $x^2 - 10x + 25 = (x - 5)(x - 5) = (x - 5)^2$

Quadratics can be rearranged to include a perfect square. This is called 'completing the square'.

Thus $x^2 - 10x + 19 = x^2 - 10x + 25 - 6 = (x - 5)^2 - 6$

$19 = 25 - 6$

This can be used to solve a quadratic equation, giving the answer in surd form.

Solving quadratic equations by using the formula Unit 3

The formula $x = \frac{-b \pm \sqrt{b^2 - 4ac}}{2a}$ is derived from completing the square. It can be used to solve any quadratic equation provided you have a calculator to work out the square root. The formula is printed on the exam paper.

a is the coefficient of x^2, b is the coefficient of x and c is the constant.

You will be told the degree of accuracy required in your answer: this is usually two decimal places.

Hint

When you are instructed to give your answer correct to a number of decimal places, you know that the quadratic expression cannot be factorised so you should use the formula.

Algebraic proof Unit 2

A proof is a series of mathematical steps that prove a particular statement is true.

You may be given algebraic expressions to work on, or you may have to decide on the symbols you are going to use in your proof. Write out each step of your working clearly: there will be marks for each step.

Example Graphs of quadratic functions Unit 3

C B

a Copy and complete the table of values for $y = 5 + 2x - x^2$

x	−2	−1	0	1	2	3	4
y		2		6	5		−3

b Draw the graph of $y = 5 + 2x - x^2$ for values of x from −2 to 4.

c Use your graph to solve the equation $5 + 2x - x^2 = 3$

Solution

a When $x = -2$, $y = 5 + (-4) - (4) = -3$

When $x = 0$, $y = 5$

When $x = 3$, $y = 5 + 6 - 9 = 2$

x	−2	−1	0	1	2	3	4
y	−3	2	5	6	5	2	−3

AQA Examiner's tip

In this table of values, you can see symmetry in the bottom line. Symmetry will not always appear in the table, but look out for it as a way of checking you have the correct y-values.

b Plot the points from the table and join them with a smooth curve.

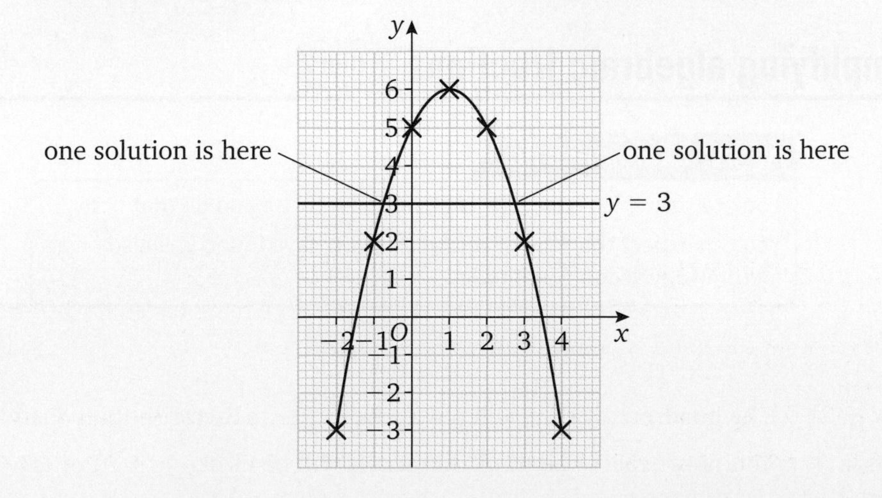

c To solve the equation $5 + 2x - x^2 = 3$ compare it with the equation for the graph: $y = 5 + 2x - x^2$

3 replaces y in equation

y has been replaced with 3 so the solutions are found where the curve crosses $y = 3$

The solutions are $x = -0.7$ and $x = 2.7$

Example Factorising quadratic expressions

Units 2 3

B A

Factorise:

a $x^2 + 5x - 24$ **b** $y^2 - 1$ **c** $3w^2 + w - 4$

Factorise completely:

d $5t^2 + 15t + 10$

Solution

a $x^2 + 5x - 24 = (x + \ldots)(x - \ldots)$

There are many pairs of factors of -24.

They have to add up to $+5$ so the pair that belong here is $+8$ and -3.

$x^2 + 5x - 24 = (x + 8)(x - 3)$

The first terms in the brackets have to be x and x to multiply to x^2.

The second terms in the brackets multiply to -24 so one is + and the other is −.

AQA Examiner's tip

You should **always** check your factors by multiplying them out to see if you obtain the correct quadratic.

b $y^2 - 1 = (y + \ldots)(y - \ldots)$

The factors of -1 are $+1$ and -1.

$y^2 - 1 = (y + 1)(y - 1)$

Here the term in y has disappeared so the two gaps must be the same number with opposite signs.

This is an example of 'the difference of two squares': $y^2 - 1^2$

c $3w^2 + w - 4 = (3w\ldots)(w\ldots)$

The factors of -4 could be $+2$ and -2,
or $+1$ and -4,
or $+4$ and -1.

As the coefficient of w^2 is not 1, you cannot put the signs into the brackets at this stage.

Try out the various combinations to find the one that gives $+w$ as the middle term. FOIL is useful for this.

$3w^2 + w - 4 = (3w + 4)(w - 1)$

d $5t^2 + 15t + 10 = 5(t^2 + 3t + 2)$
$= 5(t + 1)(t + 2)$

'Factorise completely' warns you to take out an extra factor.

Example Simplifying algebraic fractions

Unit 2

A

Simplify $\dfrac{5x^2 - 11x + 2}{x^2 + 7x - 18}$

AQA Examiner's tip

Look to see which quadratic is easier to factorise and do that first.

You can expect that one of the factors you have found will also be a factor of the second quadratic.

Solution

$x^2 + 7x - 18 = (x + 9)(x - 2)$ The quadratic with x^2 will always be easier to factorise than one with $5x^2$.

$5x^2 - 11x + 2 = (5x \ldots)(x \ldots)$ You now expect one of your brackets will be either $(x + 9)$ or $(x - 2)$.

The factors of $+2$ could be $+1$ and $+2$,
or -1 and -2.

Try $(5x - 1)(x - 2)$ FOIL leads to $-11x$ as the middle term.

$$\frac{5x^2 - 11x + 2}{x^2 + 7x - 18} = \frac{(5x - 1)\cancel{(x - 2)}}{(x + 9)\cancel{(x - 2)}} = \frac{5x - 1}{x + 9}$$

Example Solving quadratic equations by using the formula (Unit 3)

A

Solve the equation $3x^2 - 5x - 4 = 0$. Give your answers correct to two decimal places.

Solution

Write down the values of a, b and c.

$a = 3, b = -5, c = -4$ — Make sure that you put in the signs.

Work out the value of $b^2 - 4ac$.

$b^2 - 4ac = (-5)^2 - 4 \times 3 \times -4 = 25 + 48 = 73$ — Take great care with the signs here.

$\sqrt{73} = 8.544...$ — If you are trying to find the square root of a negative number, you have made a mistake in the previous line.

Substitute in the formula.

$$x = \frac{-b \pm \sqrt{b^2 - 4ac}}{2a} = \frac{+5 \pm 8.544}{6}$$

Do not correct back to 2 decimal places until the end of the solution.

$x = \frac{13.544}{6} = 2.257$ — using the positive square root first

or $x = -\frac{3.544}{6} = -0.590$ — using the negative square root

Answers: $x = 2.26$ or $x = -0.59$

Example Completing the square (Unit 2)

A*

Solve the equation $x^2 + 6x + 7 = 0$ by completing the square. Give your answers in surd form.

Solution

$x^2 + 6x + 7 = (x + 3)^2 - 2$ — Halve the coefficient of x to find the number in the brackets. $3^2 = 9$ so subtract 2 to equal the constant term 7.

$(x + 3)^2 - 2 = 0$

$(x + 3)^2 = 2$ — adding 2 to both sides

$x + 3 = \pm\sqrt{2}$ — taking the square root of both sides

$x = \pm\sqrt{2} - 3$ — taking 3 from both sides

Practise... Quadratic functions (Units 2 3)

D C B A A*

D

1 Prove that $4(3x - 2) + 2(x - 10) \equiv 14(x - 2)$

AQA Examiner's tip

Work from the left-hand side until you reach the expression on the right-hand side. Do **not** work from both sides.

D C

2 **a** Copy and complete the table of values for $y = 3 - x^2$

x	−3	−2	−1	0	1	2	3
y	−6		2	3		−1	

b Draw the graph of $y = 3 - x^2$ for values of x from −3 to 3. You should work on a grid like this.

c Write down the equation of the line of symmetry of this curve.

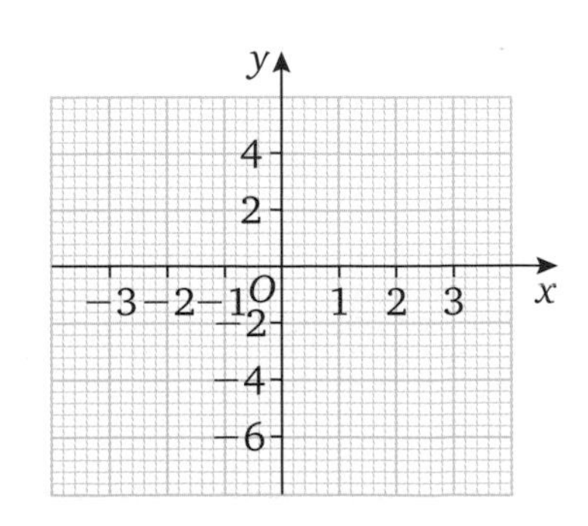

C

3 **a** Copy and complete the table of values for $y = 2x^2 - 3x - 5$

x	−3	−2	−1	0	1	2	3	4
y	22	9			−6		4	15

b Draw the graph of $y = 2x^2 - 3x - 5$ for values of x from −3 to 4.

c Use your graph to find the values of x when $y = 3$

B A

4 Factorise:

a $p^2 + 5p + 4$ **c** $30 - t - t^2$ **e** $2x^2 + 7x + 6$ **g** $5z^2 + 11z - 12$

b $q^2 - 5q - 36$ **d** $9 - v^2$ **f** $3y^2 - 8y + 5$ **h** $4w^2 - 25$

A

5 Factorise completely:

a $3x^2 - 15x + 18$ **b** $2y^2 - 8$ **c** $t^3 + 2t^2 - 24t$

B A

6 Solve these equations.

a $(a + 5)(a - 8) = 0$

b $(b + 7)(4b - 3) = 0$

c $c^2 - 7c + 12 = 0$

d $d^2 + d - 2 = 0$

e $30 - 7e - e^2 = 0$

f $2f^2 + 3f + 1 = 0$

g $3g^2 + 2g - 16 = 0$

h $4 + 4h + h^2 = 0$

i $16k^2 - 1 = 0$

j $4m^2 + m = 0$

A

7 Simplify:

a $\dfrac{(a + 3)(a - 8)}{(a - 8)(a - 3)}$ **c** $\dfrac{c^2 - 5c - 50}{c^2 + 10c + 25}$ **e** $\dfrac{e^2 - 9e + 14}{3e^2 - 19e - 14}$

b $\dfrac{b^2 - 49}{b^2 - 2b - 63}$ **d** $\dfrac{5d^2 + 13d - 6}{d^2 + d - 6}$ **f** $\dfrac{9f^2 - 1}{6f^2 - f - 1}$

8 Solve these equations.

a $\dfrac{4}{2x - 5} + \dfrac{2}{x - 1} = 2$ **b** $\dfrac{2}{y - 7} - \dfrac{1}{y - 1} = \dfrac{1}{9}$ **c** $\dfrac{4}{t + 5} + \dfrac{6}{3t - 25} = 1$

9 Solve these equations, giving your answers correct to two decimal places.

a $a^2 + 5a - 2 = 0$ **c** $2c^2 + 3c - 7 = 0$ **e** $2e^2 = 5e + 11$

b $b^2 - 7b + 8 = 0$ **d** $3d^2 - 8d + 2 = 0$ **f** $6 - 4f^2 = f$

10 Explain why you cannot solve the equation $2y^2 - 5y + 6 = 0$

11 Prove that the difference of the squares of two consecutive integers is always an odd number.

A*

12 **a** $x^2 + 8x + 5 = (x + 4)^2 - a$
Work out the value of a.

b $y^2 - 20y + 44 = (y - 10)^2 - b$
Work out the value of b.

c Solve the equation $k^2 + 12k + 5 = 0$ by completing the square.
Give your answer in surd form.

6

Higher graphs

Unit 3

21 Cubic, circular and exponential functions

Key terms

Write down definitions for the following words. Check your answers in the glossary of your Student Book.

circular functions
cubic functions
discontinuous
exponential functions
function
reciprocal

Revise... Key points

Cubic functions (Unit 3)

A **cubic function** has a term in x^3 but no higher power of x.

$y = x^3$, $y = x^3 + 5x$, $y = 2x^3 - x^2 - 1$, $y = 4x - x^3$, are examples of cubic functions.

This is a sketch graph of $y = x^3 - 3x^2 - 6x + 8$

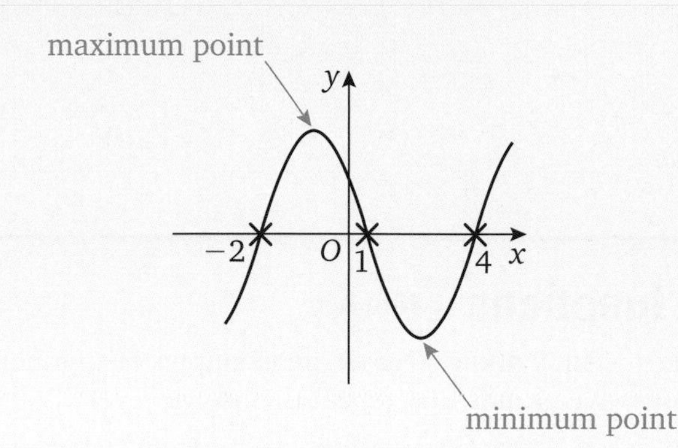

The curve crosses the x-axis three times: at $x = -2$, $x = 1$ and $x = 4$. There are two turning points on a cubic: a maximum and a minimum, but these may be less obvious than in the example above.

This is a sketch graph of $y = x^3 + 3$

This cubic crosses the x-axis once only.

At the point (0, 3), the curve flattens out and then continues with a positive gradient. It is as though the maximum and the minimum points have come together. This is called a 'point of inflexion'.

point of inflexion

(You do not have to know this term for the examination.)

The worked example on page 95 shows a cubic that crosses the x-axis once and touches it once.

If the coefficient of x^3 is negative, the graph starts in the second quadrant and finishes in the fourth quadrant.

This is a sketch graph of $y = -x^3$

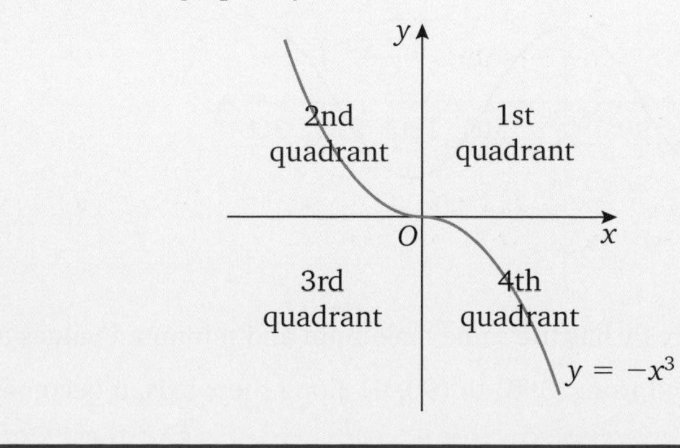

Reciprocal functions Unit 3

$y = \frac{a}{x}$, where a is a constant, is called a **reciprocal** function.

The reciprocal of x is $\frac{1}{x}$

This is a sketch graph of $y = \frac{1}{x}$

This graph is **discontinuous** as it is not one continuous curve but two separate curves. When the value of x is close to zero, the value of y is very large. When $x = 0.01$, $y = 100$, when $x = 0.001$, $y = 1000$ and when $x = 0.0001$, $y = 10\,000$, and so on.

The function is undefined at $x = 0$, so it never crosses the y-axis but gets very close to it.

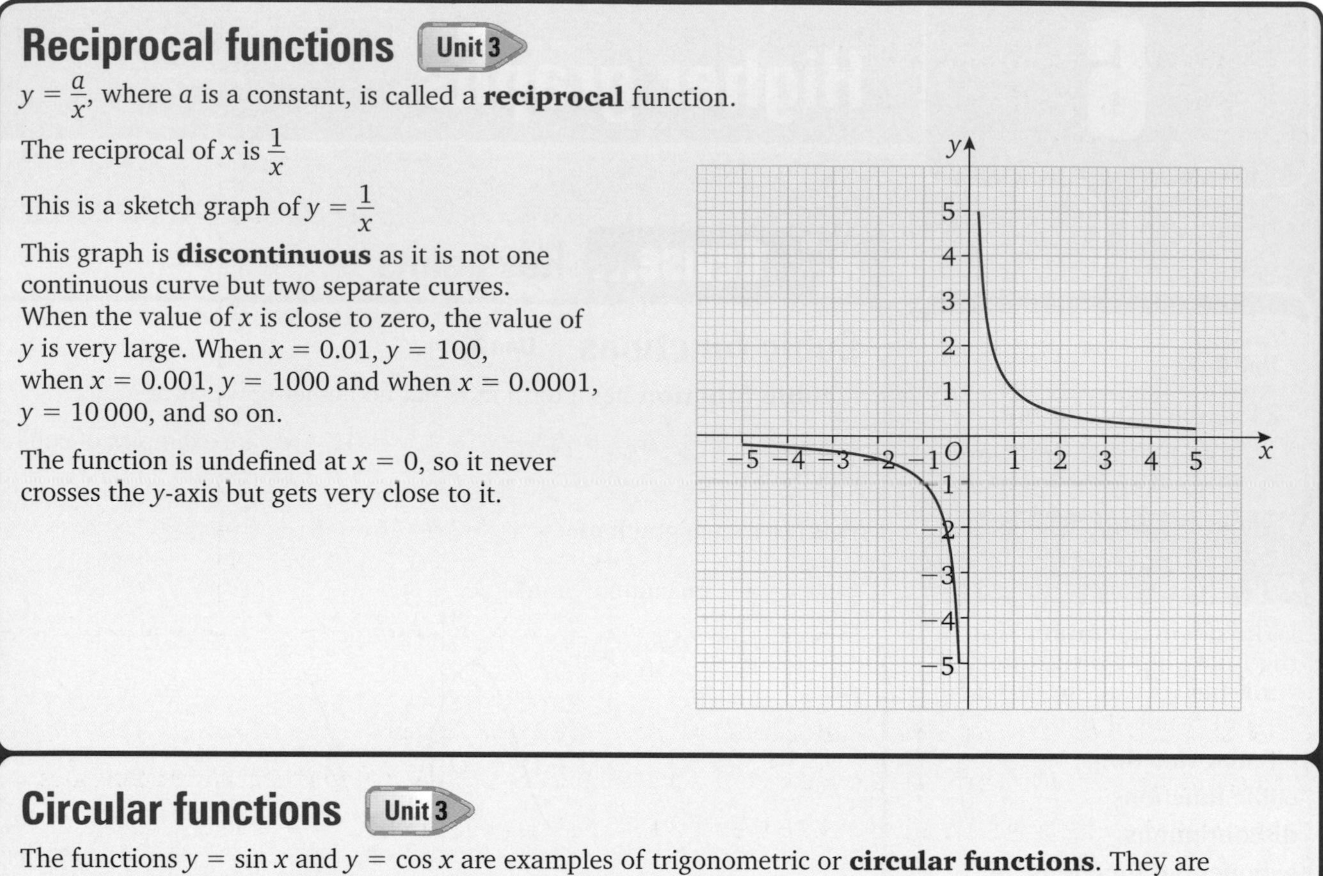

Circular functions Unit 3

The functions $y = \sin x$ and $y = \cos x$ are examples of trigonometric or **circular functions**. They are represented by wave graphs that repeat themselves every 360°.

This is the graph of $y = \sin x$. It could be extended indefinitely in either direction.

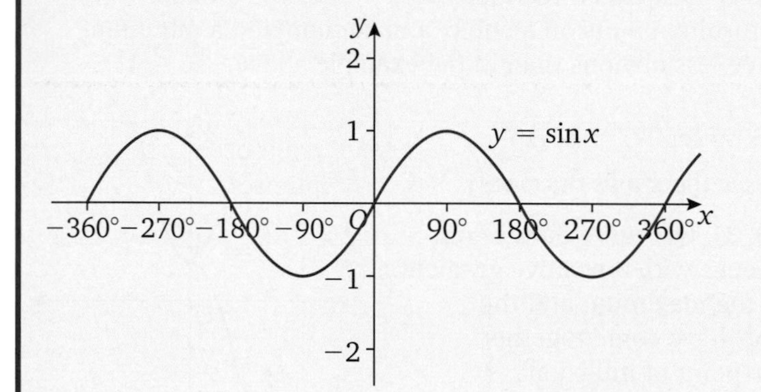

The maximum value of $\sin x$ is 1 and the minimum value is -1.

This is the graph of $y = \cos x$

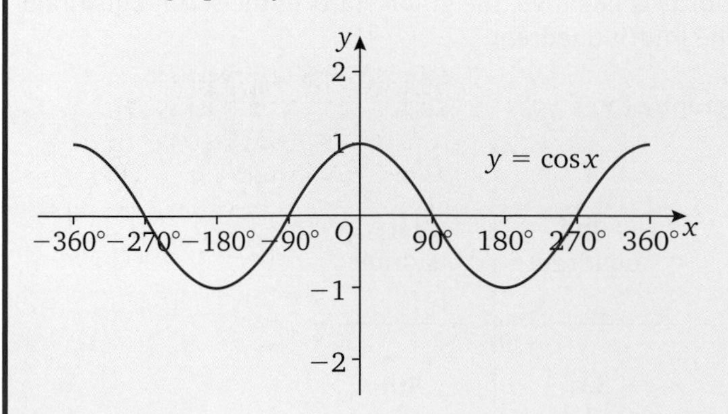

The graph of $y = \cos x$ has the same maximum and minimum values as $y = \sin x$

If $y = \sin x$ is moved from (0, 0) to (90, 0) along the x-axis, it becomes $y = \cos x$

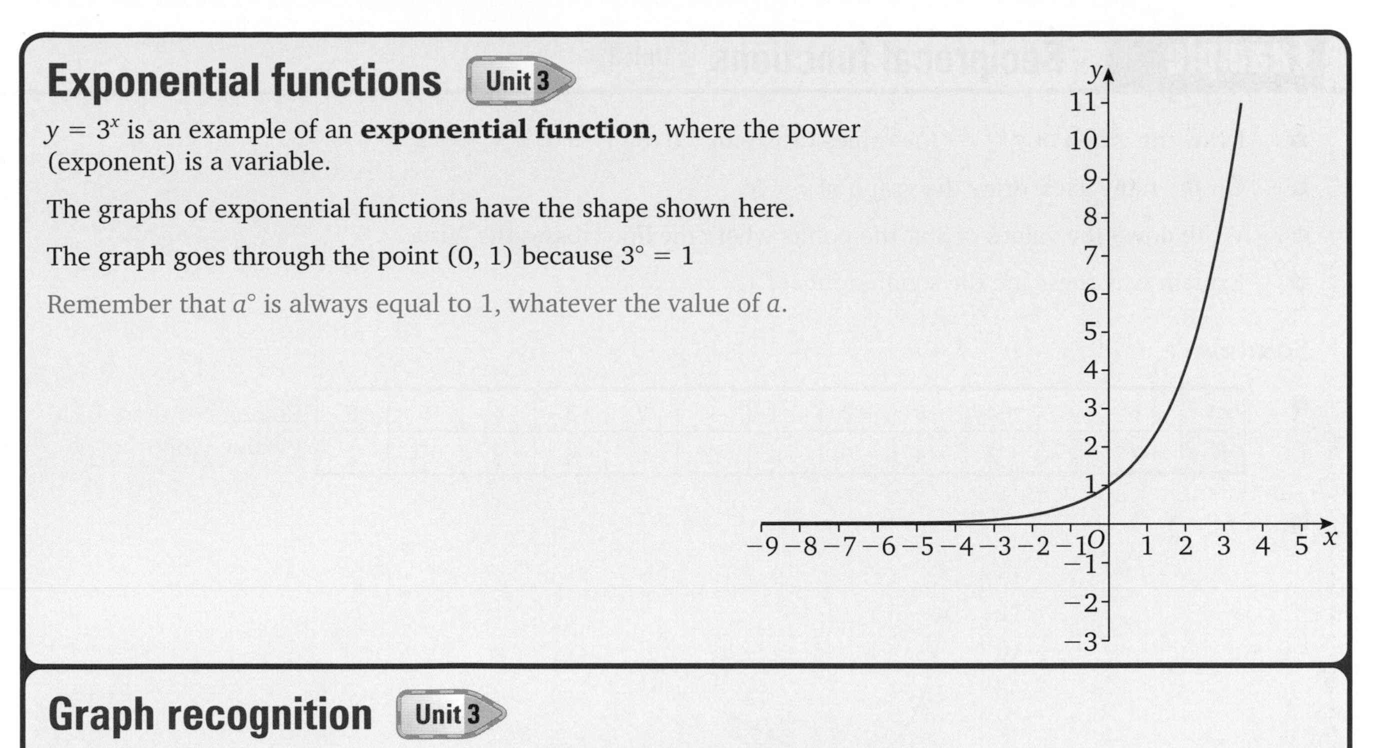

Exponential functions Unit 3

$y = 3^x$ is an example of an **exponential function**, where the power (exponent) is a variable.

The graphs of exponential functions have the shape shown here.

The graph goes through the point (0, 1) because $3^0 = 1$

Remember that a^0 is always equal to 1, whatever the value of a.

Graph recognition Unit 3

You should be able to draw and to recognise the graphs of different functions.

In addition to those described above, you should recognise linear graphs and graphs of quadratic functions.

The worked example shows how to pick up the clues that help you to do this.

Example Cubics Unit 3

B

a Draw the graph of $y = 3x^2 - x^3$ for values of x from −2 to 4.

b Use your graph to find solutions of the equation $3x^2 - x^3 = 3$

Solution

a

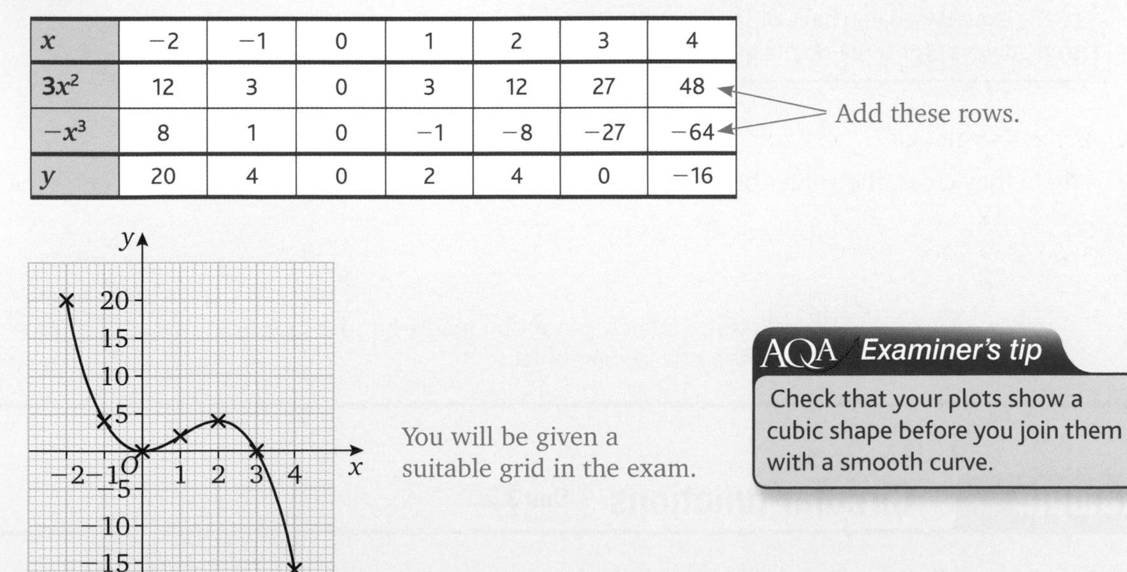

x	−2	−1	0	1	2	3	4
$3x^2$	12	3	0	3	12	27	48
$-x^3$	8	1	0	−1	−8	−27	−64
y	20	4	0	2	4	0	−16

Add these rows.

You will be given a suitable grid in the exam.

AQA Examiner's tip

Check that your plots show a cubic shape before you join them with a smooth curve.

b Read off the points where the line $y = 3$ crosses the curve.

$x = -0.9, 1.4, 2.5$ You may get one, two or three answers.

Example Reciprocal functions Unit 3

B

a Draw the graph of $y = \dfrac{12}{x}$ for values of x from −6 to 6.

b On the same axes, draw the graph of $y = x$

c Write down the values of x at the points where the line crosses the curve.

d Explain why these are the square roots of 12.

Solution

a

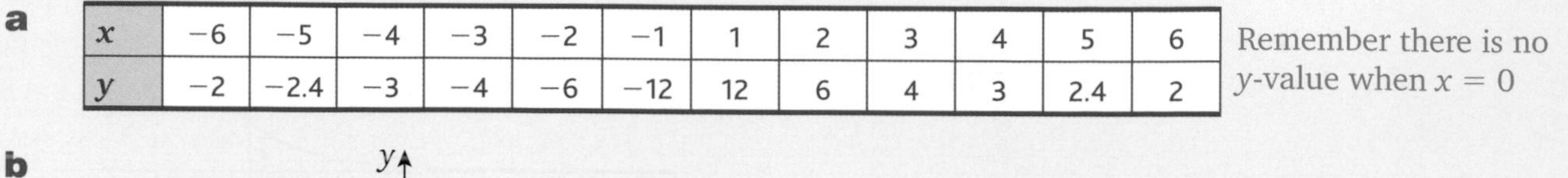

x	−6	−5	−4	−3	−2	−1	1	2	3	4	5	6
y	−2	−2.4	−3	−4	−6	−12	12	6	4	3	2.4	2

Remember there is no y-value when $x = 0$

b

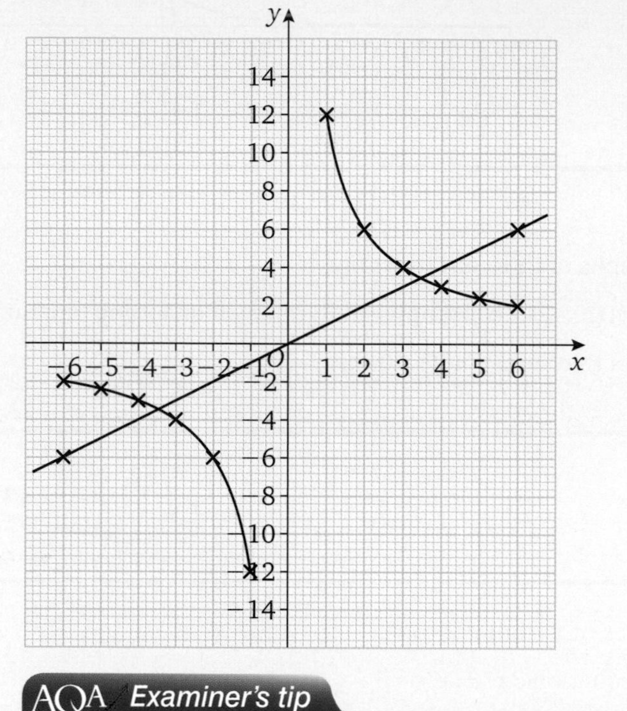

AQA Examiner's tip

Beware of thinking that the line $y = x$ is always at 45° to the axes. When you have different scales, this is not true. Always plot three points in order to draw the line.

c $x = \simeq \pm 3.5$

d Where they cross, the values of x and y satisfy both $y = x$ and $y = \dfrac{12}{x}$ simultaneous equations

So $x = \dfrac{12}{x}$ Multiply both sides by x.

$x^2 = 12$

$x = \pm\sqrt{12}$ $\sqrt{12} = 3.464\ldots$ which cannot be read from the graph to more than one decimal place.

Example Circular functions Unit 3

A A*

a Plot the graph of $y = 2\cos x$ for values of x from 0° to 360°.

b Use your graph to find the solutions of the equation $2\cos x = 1.2$ that lie between 0° and 360°.

c Write down two solutions between 360° and 720°.

Solution

a

x	0	30	60	90	120	150	180	210	240	270	300	330	360
y	2	1.7	1	0	−1	−1.7	−2	−1.7	−1	0	1	1.7	2

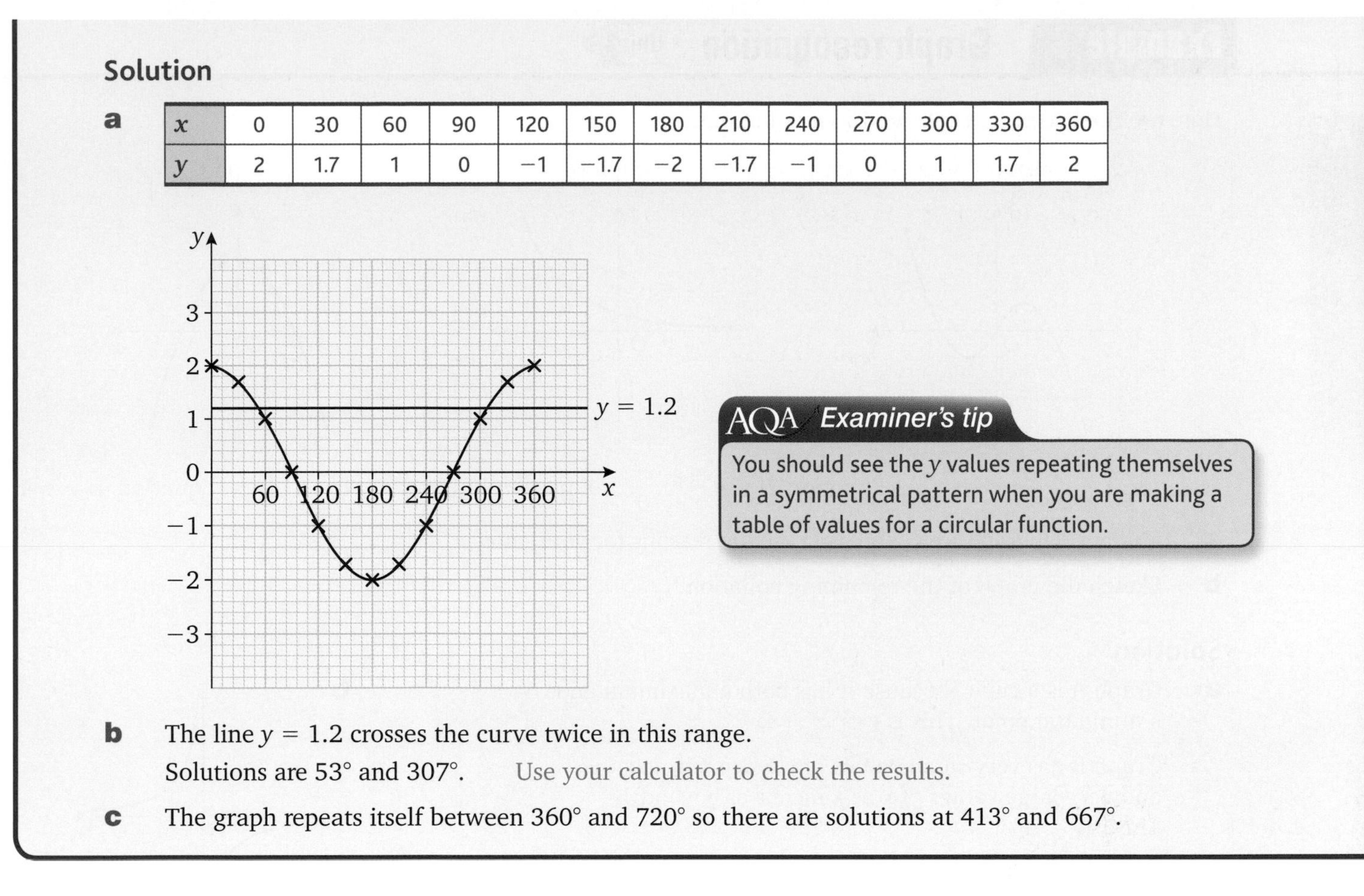

AQA Examiner's tip

You should see the y values repeating themselves in a symmetrical pattern when you are making a table of values for a circular function.

b The line $y = 1.2$ crosses the curve twice in this range.
Solutions are 53° and 307°. Use your calculator to check the results.

c The graph repeats itself between 360° and 720° so there are solutions at 413° and 667°.

Example Exponential functions Unit 3

A*

a Draw the graph of $y = 3^x$ for values of x from −3 to 3.

b Use your graph to solve the equation $3^x = 8$

Solution

a

x	−3	−2	−1	0	1	2	3
y	0.03	0.11	0.33	1	3	9	27

When x is negative, the values of y are all less than 1. For large negative values of x the y values become very close to zero but never actually reach it.

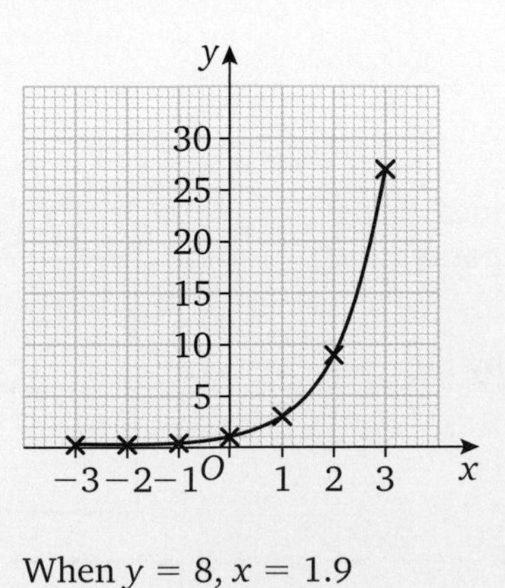

b When $y = 8$, $x = 1.9$

Aim higher

Working with the graphs of circular functions or exponential functions, you are working at Grade A/A*.

Example Graph recognition Unit 3

A*

Here are three sketch graphs and four equations.

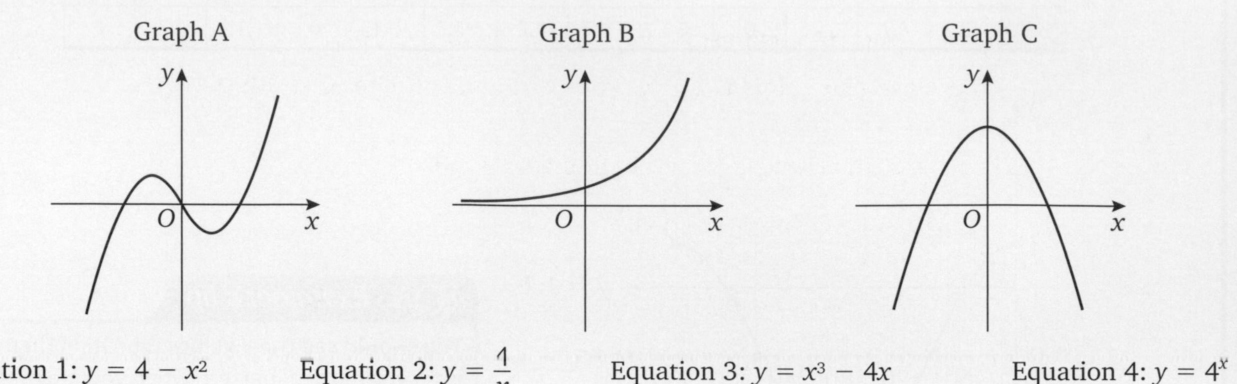

Equation 1: $y = 4 - x^2$ Equation 2: $y = \frac{1}{x}$ Equation 3: $y = x^3 - 4x$ Equation 4: $y = 4^x$

a Match each graph to its equation, giving reasons for your choice.

b Sketch the graph of the remaining equation.

Solution

a Graph A is a cubic, because it has both a maximum and a minimum point. This is $y = x^3 - 4x$

Graph B has very small values of y when x is negative and very large values of y as x increases beyond 1. This is $y = 4^x$

Graph C is a parabola, so this must be a quadratic. It is $y = 4 - x^2$

b

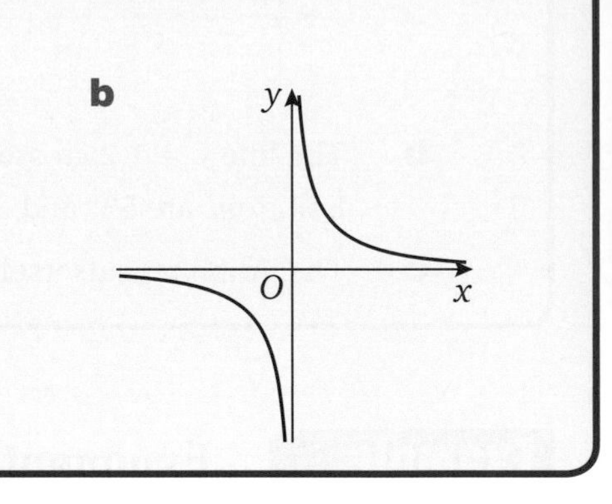

Practise... Higher graphs Unit 3

B A A*

B

1 **a** Copy and complete the table of values for $y = x^3 - 3x + 5$

x	−3	−2	−1	0	1	2	3
y	−13		7		3		23

b Draw the graph of $y = x^3 - 3x + 5$ for values of x from −3 to 3.

c Use your graph to solve the equations.

i $x^3 - 3x + 5 = 0$ **ii** $x^3 - 3x + 1 = 0$

2 A rectangular box is constructed from a square of card measuring 20 cm by 20 cm. A square of side x cm is cut from each corner and the flaps are folded up to make the box.

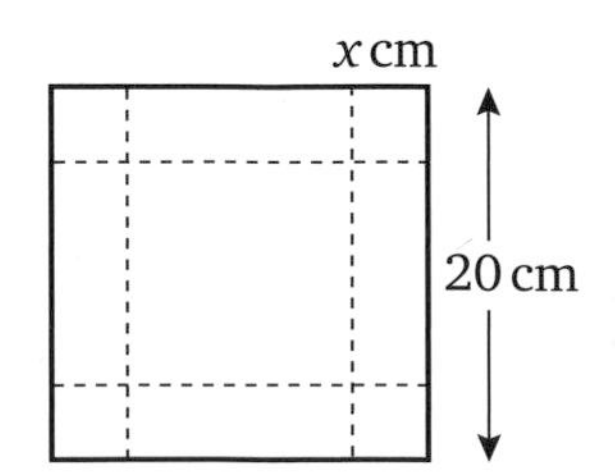

a Show that the volume of the box, V cm³, is given by the formula $V = x(20 - 2x)^2$

b Explain why the volume is zero when $x = 0$ or 10.

c Copy and complete the table of values for $V = x(20 - 2x)^2$

x	0	1	2	3	4	5	6	7	8	9	10
V	0		512	588			384	252			0

d Draw the graph of $V = x(20 - 2x)^2$ for values of x from 0 to 10.

e Use your graph to find the maximum possible value of the volume and the value of x that gives this volume.

3 A rectangle has an area of 45 cm^2.

The length of the rectangle is x cm and the width is y cm.

a Write down an equation connecting x and y.

b Draw a graph of y against x for values of x from 1 to 10.

c Use your graph to find the length of the rectangle when its width is 8 cm.

B

4 **a** Draw the graph of $y = \frac{1}{x}$ for values of x from -6 to 6.

b Use the graph to solve the equation $\frac{1}{x} = x - 2$

B A

5 **a** Sketch the graph of $y = \sin x$ for values of x from 0° to 720°.

b Which of the following statements are true?

i $\sin(180 - x) = \sin x$

ii $\sin(360 - x) = \sin x$

iii $\sin(360 + x) = \sin x$

iv $\sin(720 - x) = \sin(180 + x)$

In each case, give an example to demonstrate that the statement is true or false.

A A*

6 **a** Naomi says that if you double an angle, you double the sine.
Give an example to show that this is not true.

b Milo says that as the angle gets bigger, the cosine gets smaller.

i Give an example to support Milo's argument.

ii Sketch the graph of $y = \cos x$ from 0° to 360° and hence find an example to contradict Milo's argument.

7 A culture of bacteria increases by 20% each hour.

Initially there are 500 bacteria.

a How many bacteria will there be after two hours?

b Make a table of values for the function $B = 500 \times 1.2^x$ for values of x up to 10.

c Draw the graph of $B = 500 \times 1.2^x$ for values of x from 0 to 10.

d Use your graph to find the approximate time at which the number of bacteria reaches 1600.

A*

8 **a** Anjie says this is part of the graph of $y = x^3$
Explain why Anjie is wrong.

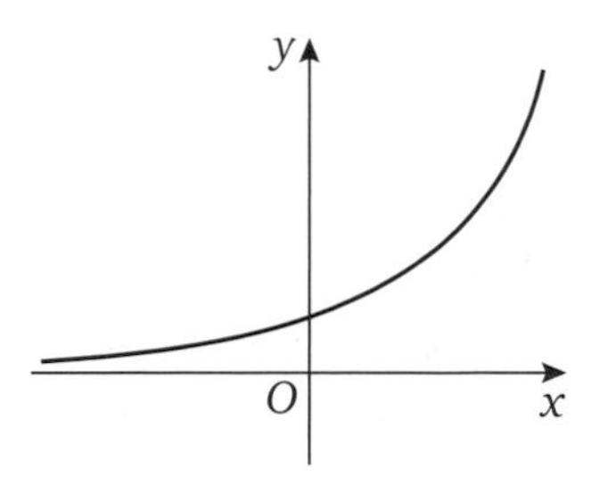

b Bindia says it is part of the graph of $y = \cos x$
Explain why Bindia is wrong.

c Cara says it is part of the graph of $xy = 5$
Explain why Cara is wrong.

d Dee says it is an exponential graph.
Explain why Dee may be correct.

Algebra

7 Inequalities and simultaneous equations

Unit 2

7 Inequalities

13 Simultaneous equations

Unit 3

13 Simultaneous equations

Key terms

Write down definitions for the following words. Check your answers in the glossary of your Student Book.

coefficient

eliminate/elimination

inequality

integer

region

simultaneous equations

substitution

variable

Revise... Key points

Inequalities (Unit 2)

Learn the four **inequality** symbols.

$<$	$\leqslant$	$>$	$\geqslant$
less than	less than or equal to	greater than	greater than or equal to

Inequalities can be represented on a number line.

This is the number line for $x \geqslant 2$

−1 0 1 2 3 4 5 x

The closed circle shows that the range includes $x = 2$

This is the number line for $x \leqslant 1$ or $x > 3$

−1 0 1 2 3 4 5 x

The open circle shows that the range does not include 3.

You may be asked to solve an inequality that looks very similar to an equation.

Follow the same process of reversing the operations.

For example: $4y + 3 < 15$

$4y < 12$ after subtracting 3 from both sides

$y < 4$ after dividing both sides by 3

You may be asked to list the **integer** values that satisfy an inequality. Remember that an integer is a whole number.

For example: List all the integer values of y such that $-10 < 3y < 6$

Divide the inequality by 3. $-3\frac{1}{3} < y < 2$

−5 −4 −3 −2 −1 0 1 2 3 y

The integer values are −3, −2, −1, 0, 1.

An inequality, such as $x + 2y > 6$, can be represented as a **region** on a graph.

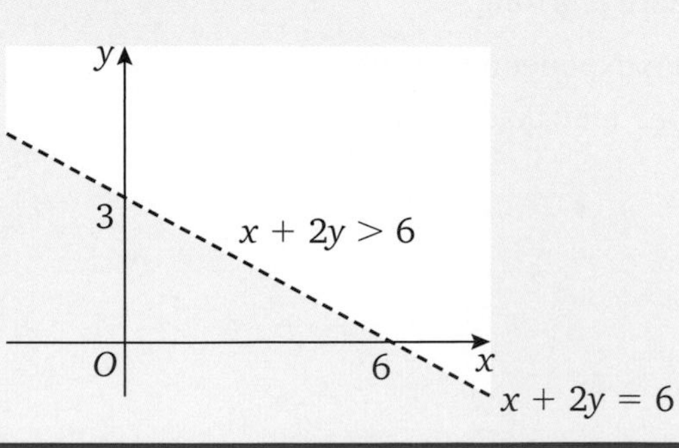

The shaded area is the region where $x + 2y > 6$

AQA Examiner's tip

Remember that if you multiply or divide by a negative number, you must reverse the inequality.

$4 < 5$ but $-4 > -5$

Hint

Drawing a number line for $-3\frac{1}{3} < y < 2$ helps you to find all the values. Remember that zero is an integer.

AQA Examiner's tip

You should always make it very clear which side of the line represents the region required.

Solving simultaneous equations Units 2 3

Simultaneous equations are two equations with two **variables** and can only be solved when reduced to one equation in one variable.

Simultaneous linear equations

Simultaneous linear equations, such as

$$3x + 2y = 4$$
$$2x - 3y = 7$$

can be solved by **elimination** of one variable or by **substitution** of a variable from one equation into the other.

Substitution is best used when one equation has x or y as the subject.

For example, $y = 2x - 5$

$$4x + y = 7$$

They can also be solved graphically. Each equation can be represented by a straight-line graph. Where the two lines cross, the values of x and y satisfy both equations so these are the solution of the simultaneous equations.

Simultaneous equations with one linear and one quadratic

There are simultaneous equations such as $5x - y = 9$

$$y = 5x^2 - 9$$

Use the linear equation to express one variable in terms of the other, and substitute this into the quadratic equation.

There will usually be two pairs of solutions to these simultaneous equations.

These can also be solved graphically by finding where the line crosses the curve.

The line might . . .

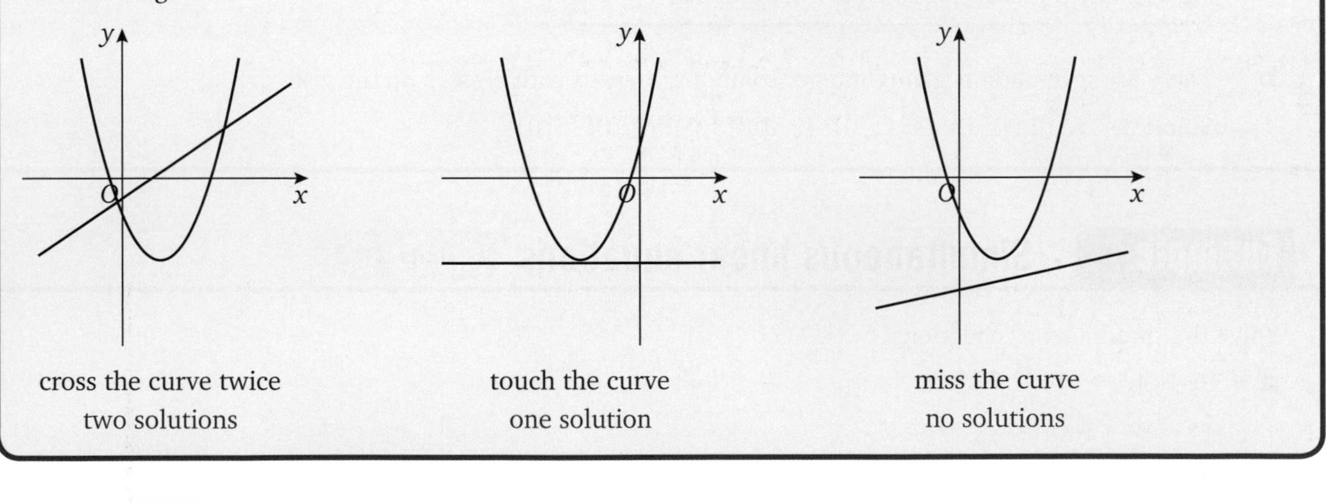

cross the curve twice — two solutions

touch the curve — one solution

miss the curve — no solutions

Example Solving inequalities Unit 2

Solve the inequality $5 - x < 8 + 2x$

Solution

$5 - x < 8 + 2x$

$5 - x - 5 < 8 + 2x - 5$ Subtract 5 from both sides.

$-x < 3 + 2x$

$-x - 2x < 3 + 2x - 2x$ Subtract $2x$ from both sides.

$-3x < 3$ Divide both sides by -3 and reverse the inequality

$x > -1$

B

Example Inequalities and regions on a graph

Unit 2

B

a On graph paper, show the region where $4x + 5y \leqslant 20$, $y < x$ and $y \geqslant 0$

b List the integer solutions of this set of inequalities.

> **AQA Examiner's tip**
>
> You should know and use this convention. 'Strict inequalities' such as $y < x$ are shown with a dashed line at the boundary and 'included inequalities' such as $y \leqslant 5$ are shown with a solid line at the boundary.

Solution

a Draw the line $4x + 5y = 20$ and the line $y = x$

x	0	5	1
y	4	0	3.2

$4x + 5y \leqslant 20$ lies below or on the line $4x + 5y = 20$

Test this with (0, 0):
$4 \times 0 + 5 \times 0 < 20$
so (0, 0) is in the region.

$y < x$ lies below the line $y = x$

Test this with (1, 2):
$2 \leqslant 1$ so (1, 2) is **not** in the region.

$y \geqslant 0$ lies on or above the x-axis.

The shaded area shows the region.

> **AQA Examiner's tip**
>
> Do not shade the region too heavily or you will not be able to find the integer points. Use hatching instead – this will show up well when your script is scanned ready for marking.

b There are seven points giving integer solutions, marked with crosses on the grid.

Solutions are (2, 1); (3, 1); (1, 0); (2, 0); (3, 0); (4, 0); (5, 0)

Example Simultaneous linear equations

B

Solve the simultaneous equations.

a $5x - 4y = 10$
$2x - 3y = 11$

b $y = 3x + 7$
$5x + 4y = 11$

Solution

a Neither the coefficients of x nor the coefficients of y are equal, so the equations have to be multiplied to get one pair of numerically equal coefficients.

$5x - 4y = 10$ Multiply by 3 to get $15x - 12y = 30$ Don't forget to multiply the right-hand side.

$2x - 3y = 11$ Multiply by 4 to get $8x - 12y = 44$ Now you have $-12y$ in both equations.

Subtract $15x - 12y = 30$ $-12y - -12y = 0$ (Signs Same Subtract)

$- \underline{8x - 12y = 44}$ Make sure that you are consistent in your subtraction: top row – bottom row.

$7x = -14$

$x = -2$

Substitute $x = -2$ into one of the original equations. Do not forget to find the second variable!

$$5 \times -2 - 4y = 10$$
$$-10 - 4y = 10$$
$$-4y = 20$$
$$y = -5$$

Check in the **other** original equation.

$$2 \times -2 - 3 \times -5 = -4 + 15 = 11 \checkmark$$

b $y = 3x + 7$

$5x + 4y = 11$

The first equation gives y in terms of x, so these equations can be solved by substitution.

$$5x + 4(3x + 7) = 11$$ This equation has only one variable.

$$5x + 12x + 28 = 11$$
$$17x + 28 = 11$$
$$17x = -17$$
$$x = -1$$

Substitute $x = -1$ into the first equation.

$$y = 3 \times -1 + 7 = 4$$

Check in the **other** original equation.

$$5 \times -1 + 4 \times 4 = -5 + 16 = 11 \checkmark$$

Example Simultaneous equations – one linear and one quadratic Unit 2

A

Solve the simultaneous equations $y + 2x = 2$
$y = 2x^2 + 5x + 5$

Solution

Rearrange the linear equation to make y the subject.

$$y = 2 - 2x$$

Substitute for y in the quadratic equation.

$$2 - 2x = 2x^2 + 5x + 5$$
$$0 = 2x^2 + 7x + 3$$
$$0 = (2x + 1)(x + 3)$$

Either $2x + 1 = 0$ or $x + 3 = 0$

$$x = -\tfrac{1}{2} \text{ or } x = -3$$

Substitute $x = -\tfrac{1}{2}$ in the linear equation:
$$y = 2 - 2 \times -\tfrac{1}{2} = 2 + 1 = 3$$

Substitute $x = -3$ in the linear equation:
$$y = 2 - 2 \times -3 = 2 + 6 = 8$$

The solutions are $x = -\tfrac{1}{2}, y = 3$ and $x = -3, y = 8$

The solutions give the coordinates of the points where the line $y + 2x = 2$ crosses the parabola $y = 2x^2 + 5x + 5$.

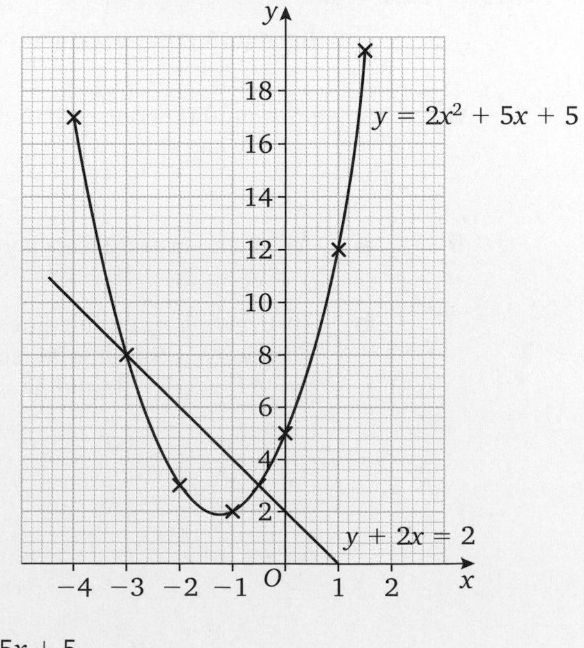

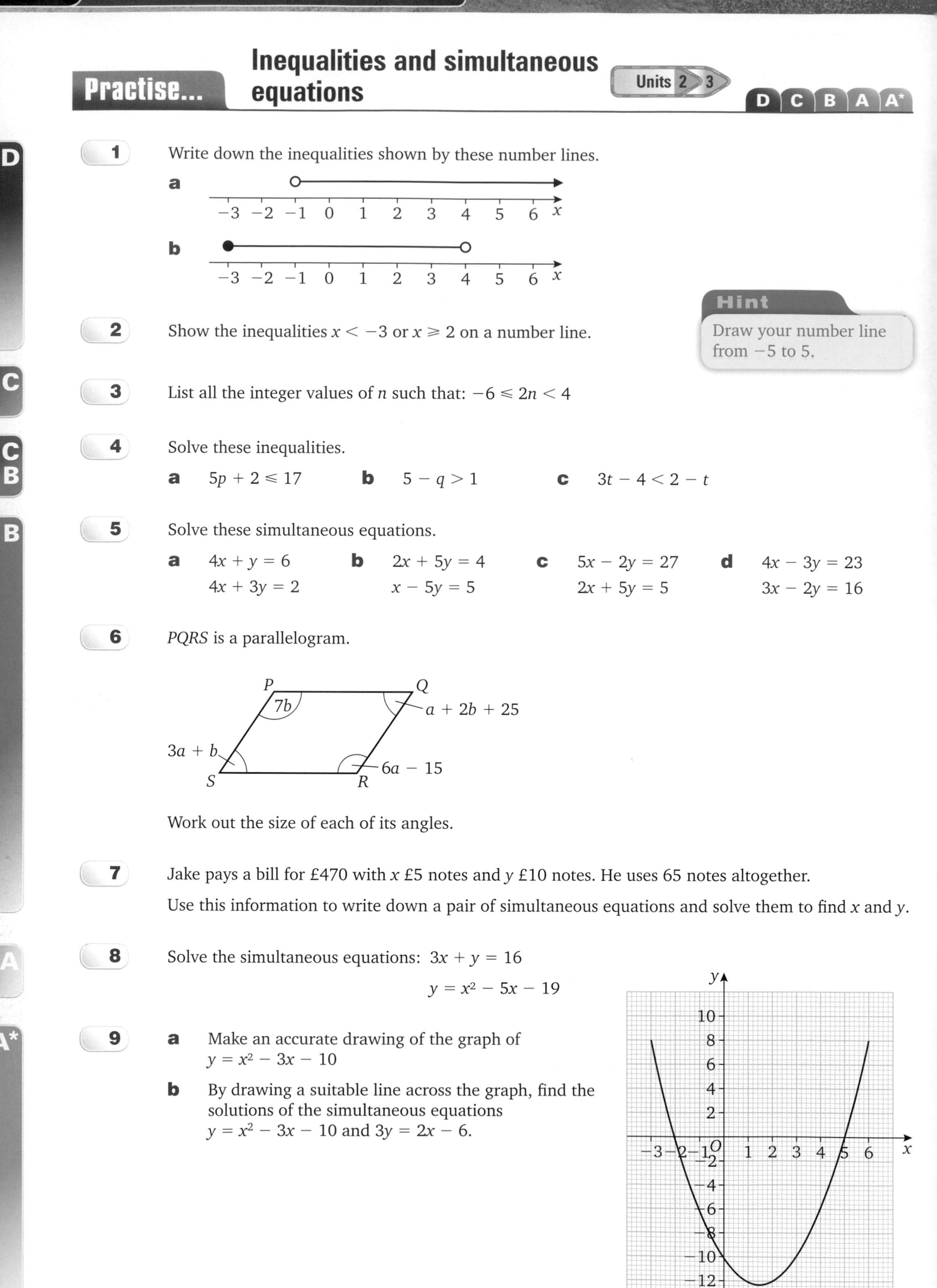

Practise... Inequalities and simultaneous equations

Units 2 3

D C B A A*

D

1 Write down the inequalities shown by these number lines.

a

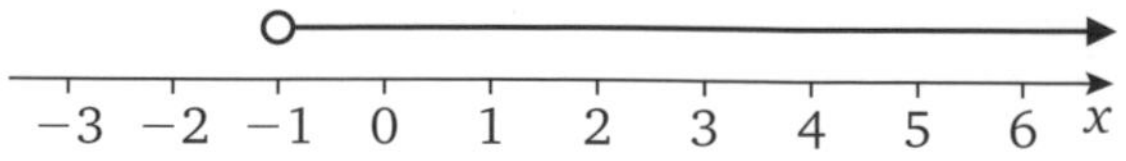

b

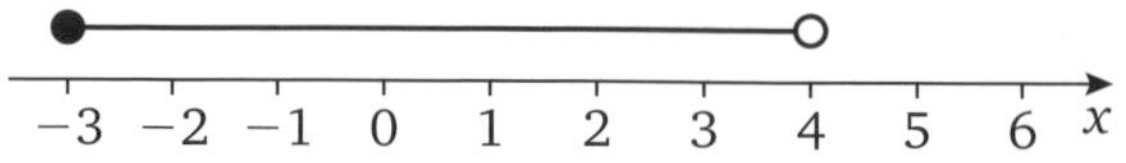

2 Show the inequalities $x < -3$ or $x \geqslant 2$ on a number line.

Hint
Draw your number line from -5 to 5.

C

3 List all the integer values of n such that: $-6 \leqslant 2n < 4$

C B

4 Solve these inequalities.

a $5p + 2 \leqslant 17$ **b** $5 - q > 1$ **c** $3t - 4 < 2 - t$

B

5 Solve these simultaneous equations.

a $4x + y = 6$
$4x + 3y = 2$

b $2x + 5y = 4$
$x - 5y = 5$

c $5x - 2y = 27$
$2x + 5y = 5$

d $4x - 3y = 23$
$3x - 2y = 16$

6 $PQRS$ is a parallelogram.

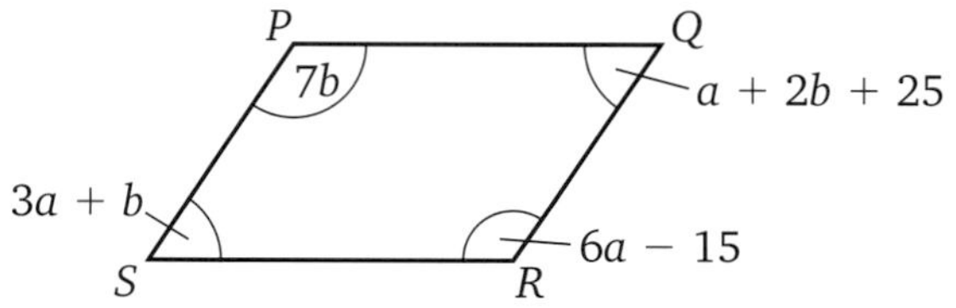

Work out the size of each of its angles.

7 Jake pays a bill for £470 with x £5 notes and y £10 notes. He uses 65 notes altogether.

Use this information to write down a pair of simultaneous equations and solve them to find x and y.

A

8 Solve the simultaneous equations: $3x + y = 16$

$y = x^2 - 5x - 19$

A*

9 **a** Make an accurate drawing of the graph of $y = x^2 - 3x - 10$

b By drawing a suitable line across the graph, find the solutions of the simultaneous equations $y = x^2 - 3x - 10$ and $3y = 2x - 6$.

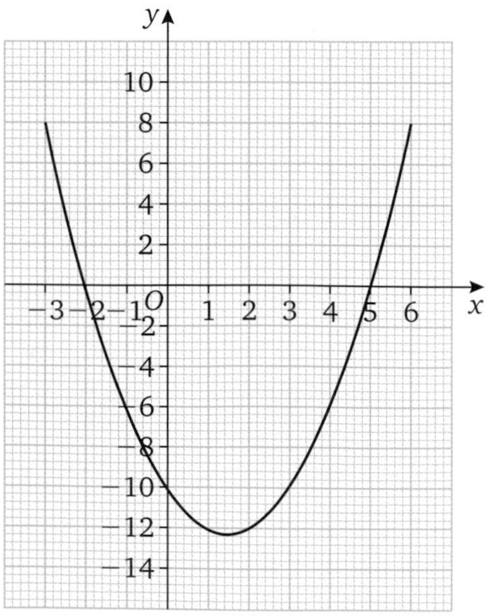

8 AQA Examination-style questions

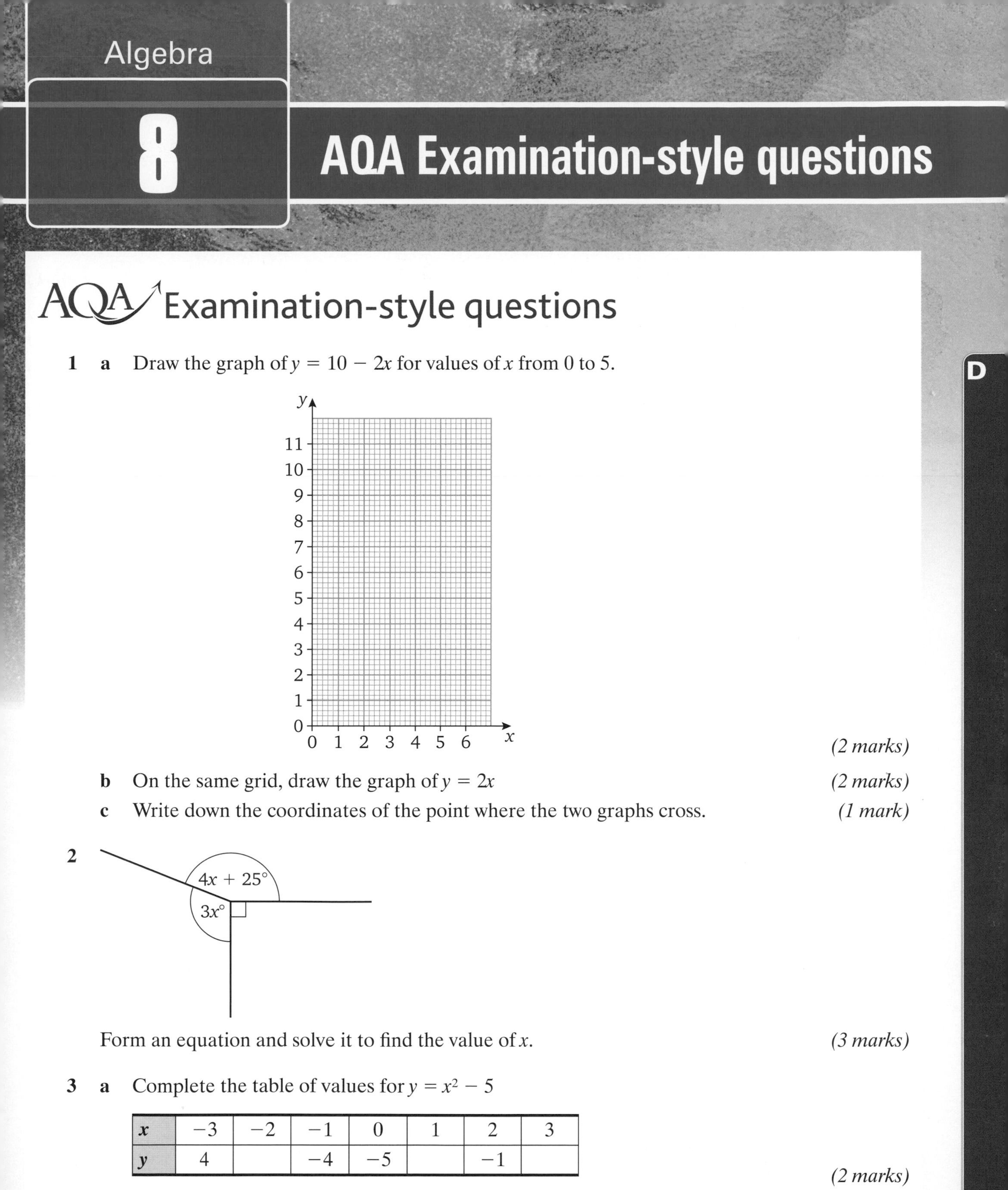

AQA Examination-style questions

D

1 **a** Draw the graph of $y = 10 - 2x$ for values of x from 0 to 5.

(2 marks)

b On the same grid, draw the graph of $y = 2x$ *(2 marks)*

c Write down the coordinates of the point where the two graphs cross. *(1 mark)*

2

Form an equation and solve it to find the value of x. *(3 marks)*

3 **a** Complete the table of values for $y = x^2 - 5$

x	−3	−2	−1	0	1	2	3
y	4		−4	−5		−1	

(2 marks)

b Draw the graph of $y = x^2 - 5$ for values of x from −3 to +3.

(2 marks)

c On the same grid, draw the graph of $y = x$ *(1 mark)*

d Hence find the solutions of the equation $x^2 - 5 = x$ *(2 marks)*

Hint

Draw the x-axis from −3 to 3.
Draw the y-axis from −5 to 4.

D

4 Factorise:

a $7x - 21$ *(1 mark)*

b $2y - y^2$ *(1 mark)*

5 Which of the words **expression**, **equation**, **formula**, **identity** describes each of the following?

a $V = \pi r^2 h$ *(1 mark)*

b $5x - 2 \equiv -2 + 5x$ *(1 mark)*

c $T = 45w + 20$ *(1 mark)*

d $4y + 3 = 7 - 2y$ *(1 mark)*

e $8 = \frac{2p - 1}{3}$ *(1 mark)*

f $p + 2q + 3r - 5t$ *(1 mark)*

D C

6 Shaun cycles from Ayton to Beeville. The graph shows his journey.

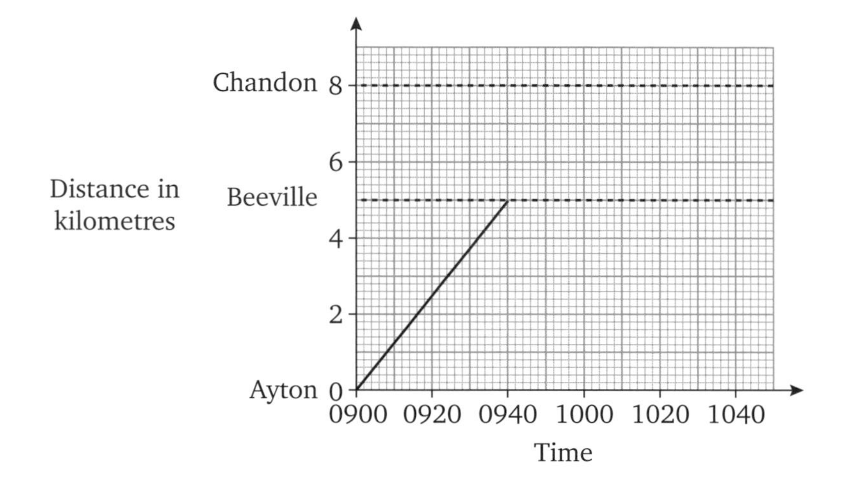

Shaun stays at Beeville for 20 minutes.

Then he cycles on to Chandon at the same speed as before.

a Copy the grid and show the rest of his journey on the grid. *(3 marks)*

b What time does he arrive at Chandon? *(1 mark)*

c Work out Shaun's average speed between Ayton and Beeville. Give your answer in km/h. *(3 marks)*

7 Solve the equations.

a $2p + 11 = 5 - p$ *(3 marks)*

b $5(3 - q) = 20$ *(3 marks)*

c $4t - 7 = 3(t + 5)$ *(3 marks)*

C

8 Kate thinks of three numbers.

Her first number is x.

Her second number is five larger than her first number.

Her third number is twice her first number.

The sum of her three numbers is 61.

a Write down an equation in x. *(2 marks)*

b Solve the equation to find the value of x. *(2 marks)*

C

9 Expand and simplify:

a $3(x - 2) - 5(1 - 2x)$ *(2 marks)*

b $4(2p - t) + 5(2t - p)$ *(2 marks)*

c $7(m - 3n) - 2(3m + 2n)$ *(2 marks)*

10 The cooking time, T minutes, for a turkey is given by the formula $T = 30W + 15$, where W is the weight in kilograms. Make W the subject of the formula. *(2 marks)*

11 a List all the integer solutions of the inequality $-3 < 2n \leqslant 6$ *(3 marks)*

b Solve the inequality $5y - 3 > 9$ *(2 marks)*

12 Here is a sequence of rectangular patterns.

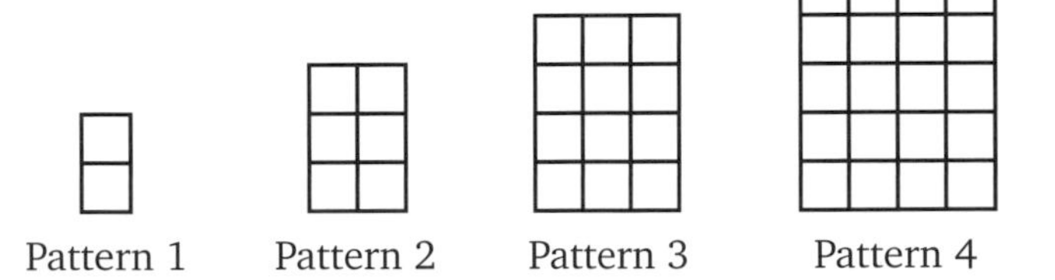

a Explain why Pattern 15 will contain 240 small squares. *(2 marks)*

b Write down an expression for the number of small squares in Pattern n. *(2 marks)*

13 A solution of the equation $x^3 + 2x = 39$ lies between $x = 3$ and $x = 4$

Use trial and improvement to find this solution, correct to one decimal place.

You must show all your trials. *(3 marks)*

C
B

14 The graph of $y = x^2 - 3x - 10$ is sketched below.

It crosses the axes at the points A, B and C.

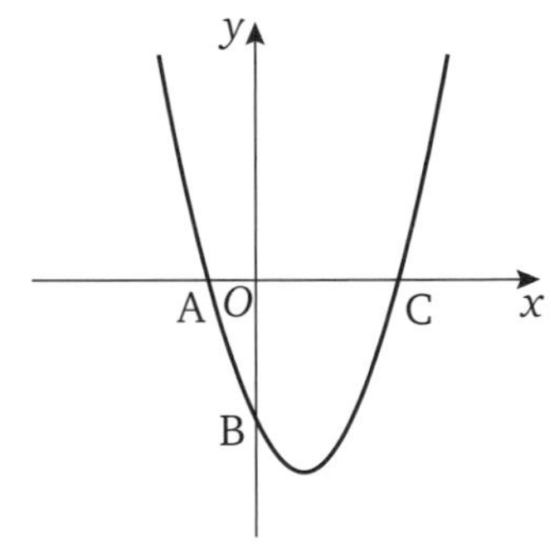

Find the coordinates of each of the points A, B and C. *(5 marks)*

B

15 $7(x - 3) - 2(5 - x) = a(x - 4) + 5$

Find the value of a. *(3 marks)*

16 a A straight line has gradient 3 and passes through the point (1, 7).
Write down the equation of the line. *(1 mark)*

b A different line has gradient $-\frac{1}{4}$ and passes through the point (2, −3).
Show that this line has the equation $x + 4y + 10 = 0$ *(2 marks)*

c Find the coordinates of the point where the two lines cross. *(3 marks)*

B

17

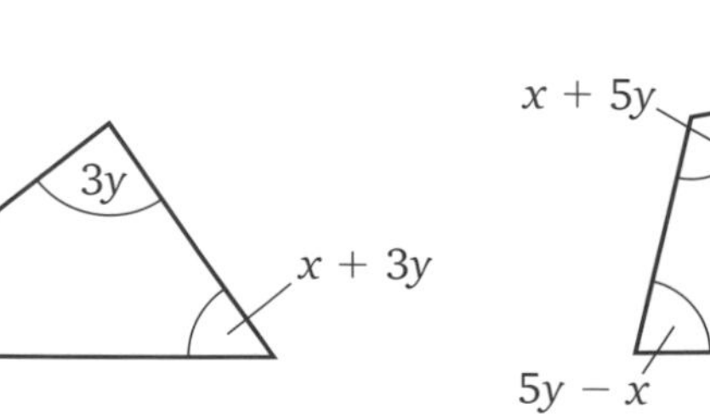

Find the values of x and y. *(4 marks)*

B A

18 Factorise:

a $p^2 - 16$ *(1 mark)*

b $q^2 + 3q - 18$ *(2 marks)*

c $2t^2 - 7t + 3$ *(2 marks)*

19 The hypotenuse of triangle PQR is x cm.
The other sides are $(x - 3)$ cm and $(x - 8)$ cm.

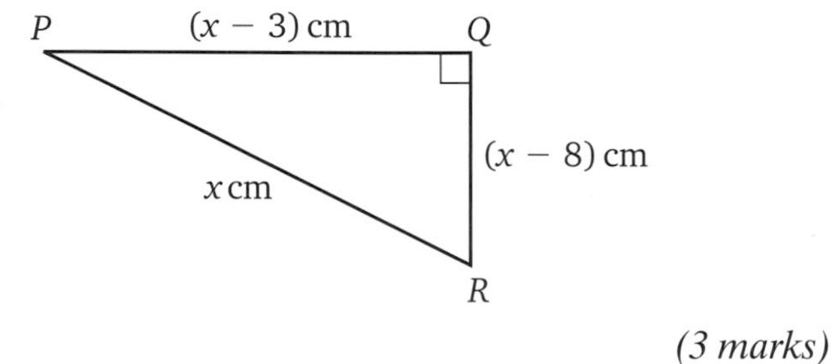

a Show that $x^2 - 22x + 73 = 0$ *(3 marks)*

b Solve this equation to find the length of the hypotenuse.
Give your answer to one decimal place. *(3 marks)*

A

20

1	2	3	4	5	6	7	8
9	10	11	12	13	14	15	16
17	18	19	20	21	22	23	24
25	26	27	28	29	30	31	32
33	34	35	36	37	38	39	40
41	42	43	44	45	46	47	48

Part of an 8-by-8 grid is shown. The shaded shape is called S_{19} because the left-hand cell in the top row contains the number 19. The sum of the numbers in S_{19} is 92.
Prove that the sum of the numbers in S_n is always a multiple of 4. *(4 marks)*

21 Simplify $\dfrac{6x^2 + 13x - 5}{9x^2 - 6x + 1}$ *(4 marks)*

A*

22 Solve the equation $\dfrac{8}{y + 1} + \dfrac{6}{y - 3} = 1$ *(5 marks)*

23 $x^2 - 8x + p = (x - q)^2 - 5$
Find the values of p and q. *(3 marks)*

24 Solve the simultaneous equations

$$y = 5x^2$$
$$2y + 9x = 22$$

(5 marks)

Geometry and measure

1 Area and volume

Unit 3

Key terms

Write down definitions for the following words. Check your answers in the glossary of your Student Book.

arc (of a circle)
base
cross-section
cylinder
frustum (of a cone)
net
perpendicular height
prism
sector
surface area
triangular prism
volume

Revise... Key points

Area of quadrilaterals and triangles (Unit 3)

The area of a rectangle, parallelogram and trapezium are all calculated by similar formulae.

Rectangle:
Area = length × width

$A = lw$

Parallelogram:
Area = base × height

$A = bh$

Trapezium:
Area = average width × height

$A = \frac{1}{2}(a + b)h$

AQA Examiner's tip

The formula for the area of a trapezium is given on the exam paper, so you do not need to learn it. But you must learn the others.

A triangle is half a parallelogram.

So the area of a triangle is $\frac{1}{2} \times b \times h$ or $A = \frac{1}{2}bh$

AQA Examiner's tip

Remember in all these formulae, the height is **perpendicular** to the base.

Circles (Unit 3)

Circumference = π × diameter, or $C = \pi d$

Area = π × (radius)^2, or $A = \pi r^2$

Diameter = 2 × radius, or $d = 2r$

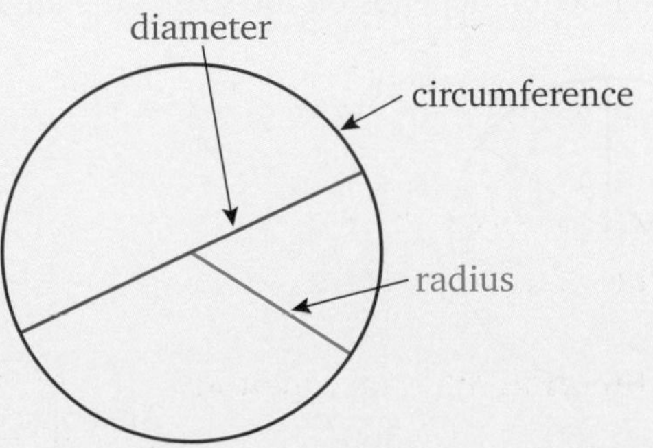

Prisms Unit 3

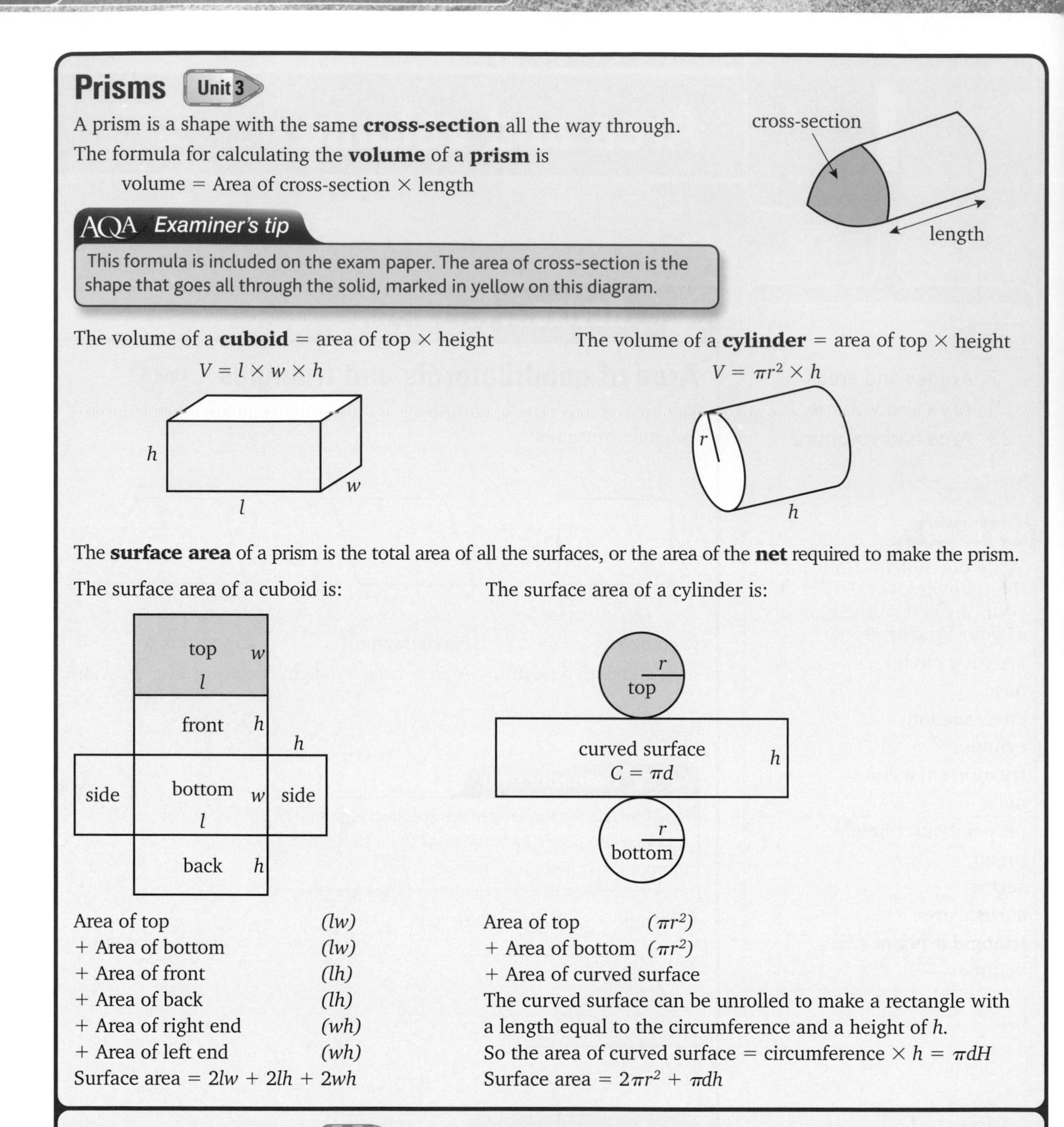

A prism is a shape with the same **cross-section** all the way through.

The formula for calculating the **volume** of a **prism** is

volume = Area of cross-section × length

AQA Examiner's tip

This formula is included on the exam paper. The area of cross-section is the shape that goes all through the solid, marked in yellow on this diagram.

The volume of a **cuboid** = area of top × height

$V = l \times w \times h$

The volume of a **cylinder** = area of top × height

$V = \pi r^2 \times h$

The **surface area** of a prism is the total area of all the surfaces, or the area of the **net** required to make the prism.

The surface area of a cuboid is:

Area of top (lw)
+ Area of bottom (lw)
+ Area of front (lh)
+ Area of back (lh)
+ Area of right end (wh)
+ Area of left end (wh)
Surface area = $2lw + 2lh + 2wh$

The surface area of a cylinder is:

Area of top (πr^2)
+ Area of bottom (πr^2)
+ Area of curved surface
The curved surface can be unrolled to make a rectangle with a length equal to the circumference and a height of h.
So the area of curved surface = circumference × h = πdH
Surface area = $2\pi r^2 + \pi dh$

Arcs and sectors Unit 3

An **arc** is part of the circumference of a circle.

A **sector** is part of the area of the circle, bounded by an arc and two radii.

The arc is a fraction of the circumference, and the sector is a fraction of the area of the circle.

The size of the fraction depends on the size of the angle at the centre.

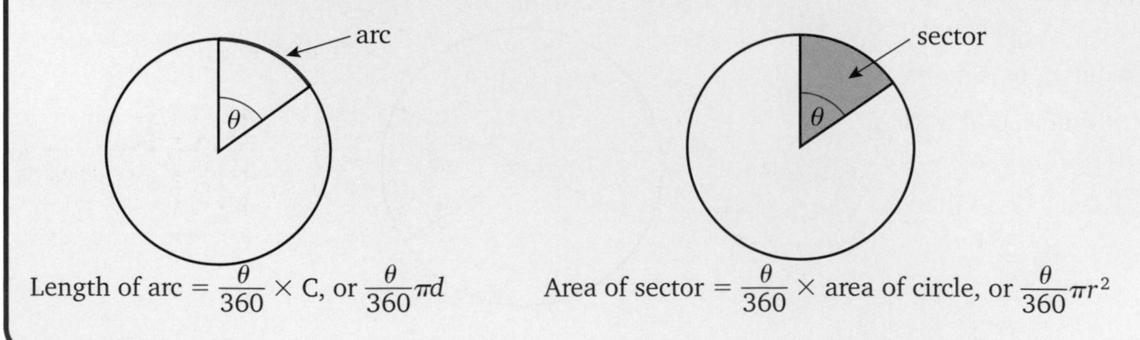

Length of arc = $\frac{\theta}{360} \times C$, or $\frac{\theta}{360}\pi d$

Area of sector = $\frac{\theta}{360} \times$ area of circle, or $\frac{\theta}{360}\pi r^2$

Pyramids, spheres, cones and frustums Unit 3

Any pyramid fits inside a prism with the same base as its cross-section and the same height.

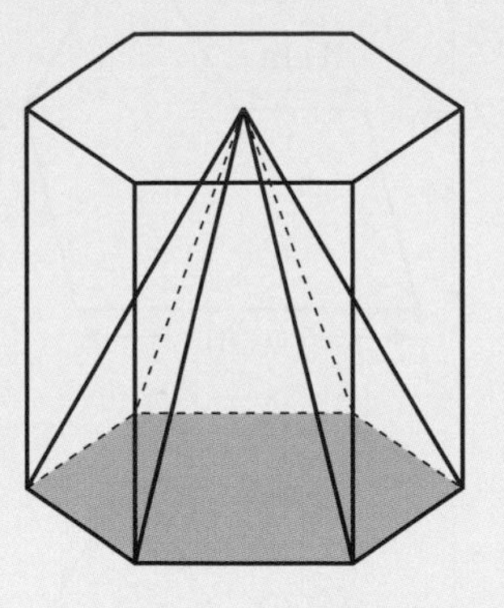

The volume of the pyramid is one-third of the volume of the prism.

The volume of a prism is area of base × height

So the volume of the pyramid is $\frac{1}{3}$ × area of base × height

AQA Examiner's tip

A prism has uniform cross-section, so $V = A \times h$

For solids that come to a vertex (cone and pyramid), $V = \frac{1}{3}A \times h$

A cone is a pyramid with a circular base, so

volume $= \frac{1}{3}$ of the volume of a cylinder

or $V = \frac{1}{3}\pi r^2 h$

The area of the curved surface $= \pi rl$

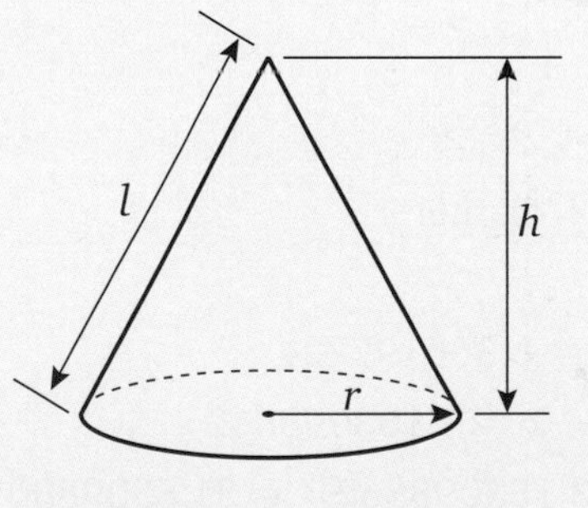

AQA Examiner's tip

These formulae are given on the exam paper.

The volume of a sphere is given by the formula $\frac{4}{3}\pi r^3$

The surface area of a sphere is given by the formula $4\pi r^2$

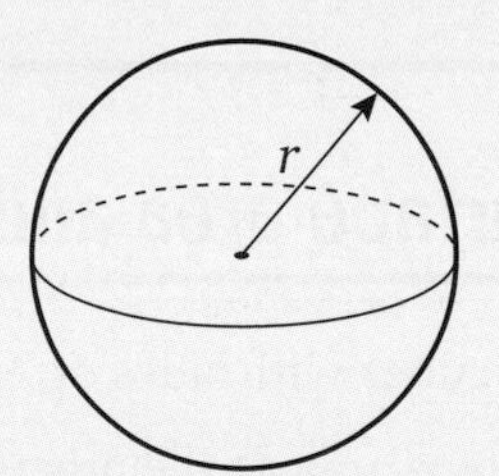

AQA Examiner's tip

These formulae are given on the exam paper.

A **frustum** of a cone is the part left when the top of a cone is removed.

You find the volume by subtracting the volume removed from the volume of the original cone.

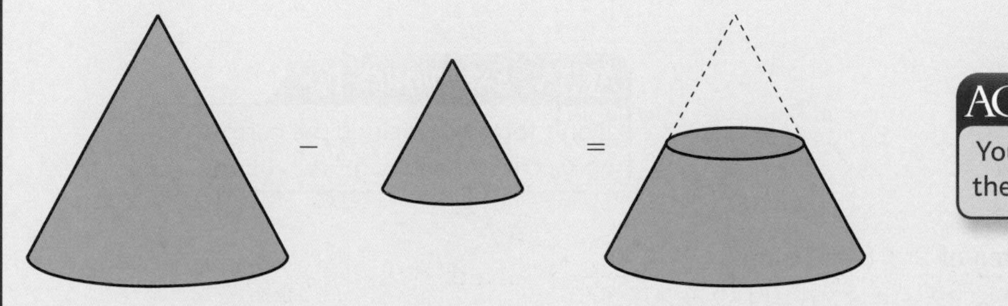

AQA Examiner's tip

You can use similar triangles to find the volume of a frustum.

Example Area of shapes made from quadrilaterals and triangles Unit 3

D

Calculate the area of this shape.

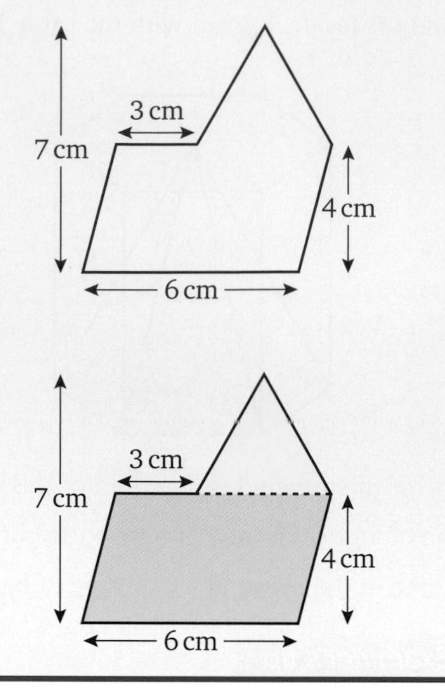

AQA Examiner's tip

Don't be afraid to draw on the diagrams on the exam paper. Annotate them with any measurements you work out.

Solution

The shape can be split into a parallelogram and a triangle.

Area of parallelogram $= lw = 6\,\text{cm} \times 4\,\text{cm} = 24\,\text{cm}^2$

Base of triangle $= 6\,\text{cm} - 3\,\text{cm} = 3\,\text{cm}$

Height of triangle $= 7\,\text{cm} - 4\,\text{cm} = 3\,\text{cm}$

Area of triangle $= \frac{1}{2}bh = \frac{1}{2} \times 3 \times 3 = 4.5\,\text{cm}^2$

Total area $= 24\,\text{cm}^2 + 4.5\,\text{cm}^2 = 28.5\,\text{cm}^2$

Example Area and circumference of circles Unit 3

D

A circle has a circumference of 20 cm.

Calculate its area.

Solution

Circumference $= \pi \times$ diameter

$20 = \pi \times$ diameter

Diameter $= 20 \div \pi = 6.366197724$

Radius $=$ diameter $\div 2 = 3.183098862$

Area $= \pi r^2 = \pi \times (3.183098862)^2 = 31.8309886$

Area $= 31.8\,\text{cm}^2$ (to 1 decimal place)

Example Surface area and volume Unit 3

D

This prism is made from cubes with sides of 2 cm.

Find the volume and surface area of the prism.

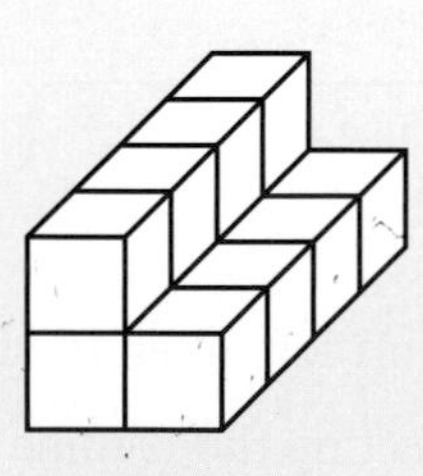

Solution

Volume:

Area of front $= 3 \times 2^2 = 12\,\text{cm}^2$

Volume $=$ area $\times$ length $= 12 \times 8 = 96\,\text{cm}^3$

Surface area:

Area of front $= 12\,\text{cm}^2$

Area of back $= 12\,\text{cm}^2$

Area of bottom $= 4 \times 8 = 32\,\text{cm}^2$

Area of left side $= 32\,\text{cm}^2$

Four other faces each have an area of $2 \times 8 = 16\,\text{cm}^2$

Total surface area $= 2 \times 12 + 2 \times 32 + 4 \times 16 = 152\,\text{cm}^2$

AQA Examiner's tip

Don't forget to include the correct units; cm^2 for area, cm^3 for volume.

Example Sector and arc Unit 3

Find the area and perimeter of the sector.

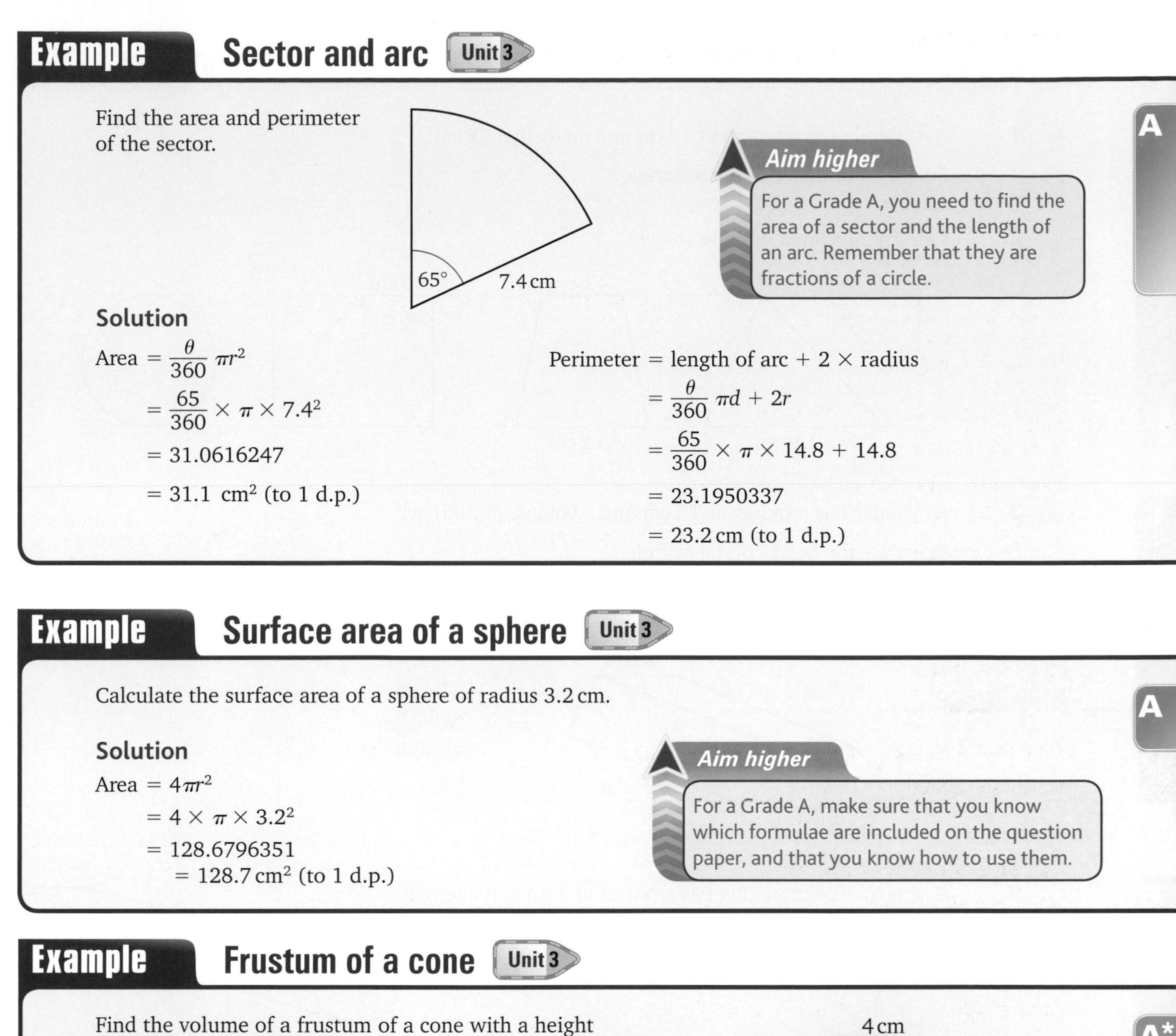

Aim higher

For a Grade A, you need to find the area of a sector and the length of an arc. Remember that they are fractions of a circle.

A

Solution

Area $= \dfrac{\theta}{360}\pi r^2$

$= \dfrac{65}{360} \times \pi \times 7.4^2$

$= 31.0616247$

$= 31.1\ \text{cm}^2$ (to 1 d.p.)

Perimeter = length of arc + 2 × radius

$= \dfrac{\theta}{360}\pi d + 2r$

$= \dfrac{65}{360} \times \pi \times 14.8 + 14.8$

$= 23.1950337$

$= 23.2$ cm (to 1 d.p.)

Example Surface area of a sphere Unit 3

Calculate the surface area of a sphere of radius 3.2 cm.

A

Solution

Area $= 4\pi r^2$

$= 4 \times \pi \times 3.2^2$

$= 128.6796351$

$= 128.7\ \text{cm}^2$ (to 1 d.p.)

Aim higher

For a Grade A, make sure that you know which formulae are included on the question paper, and that you know how to use them.

Example Frustum of a cone Unit 3

Find the volume of a frustum of a cone with a height of 9 cm, a base radius of 6 cm and a top radius of 4 cm. Give your answer in terms of π.

A*

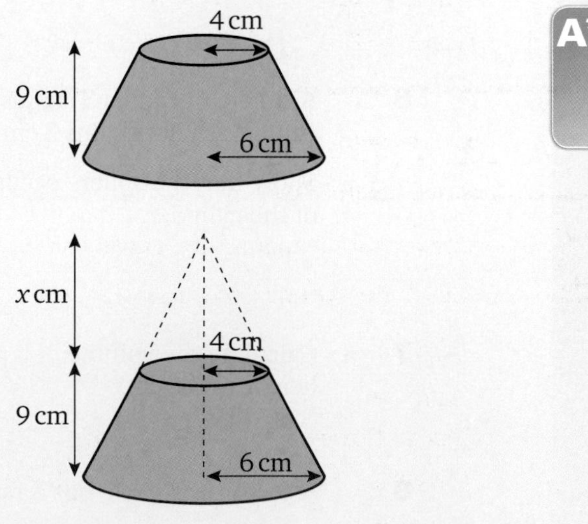

Solution

Sketch the whole cone; label the height removed x.

By similar triangles,

$\dfrac{x+9}{x} = \dfrac{6}{4}$

$\dfrac{x+9}{x} = 1.5$

$x + 9 = 1.5x$

$9 = 0.5x$

$x = 18$

So the height of the original cone = 18 + 9 = 27 cm

Original cone volume $= \frac{1}{3}\pi r^2 h$

$= \frac{1}{3} \times \pi \times 6^2 \times 27 = 324\pi\ \text{cm}^3$

Volume removed $= \frac{1}{3}\pi r^2 h$

$= \frac{1}{3} \times \pi \times 4^2 \times 18 = 96\pi\ \text{cm}^3$

Volume of frustum $= 324\pi - 96\pi = 228\pi\ \text{cm}^3$

Practise... 1 Area and volume (Unit 3)

D C B A A*

D

1 A rectangle has a length of 12 cm and an area of 36 cm^2.

Find the perimeter of the rectangle.

2 Calculate the area of these shapes:

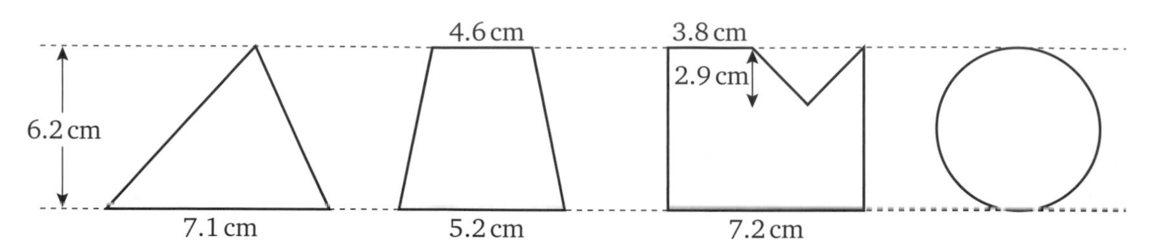

C

3 A cylinder has a radius of 4.2 cm and a volume of 200 cm^3.

Calculate the height of the cylinder.

4 Calculate the surface area and volume of this prism.

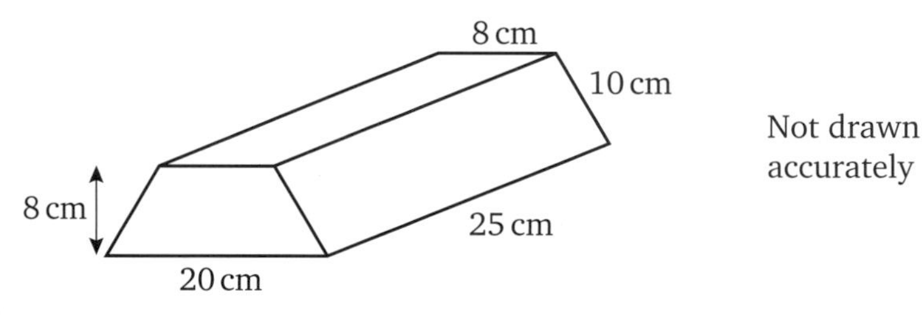

Not drawn accurately

B

5 A semicircular moulding has a radius of 4 cm and a length of 60 cm.

Calculate the volume and surface area of the moulding.

A

6 A sphere of radius 3 cm just fits inside a cylindrical container with radius 3 cm and height 6 cm.

Show that the sphere occupies exactly $\frac{2}{3}$ of the volume of the cylinder.

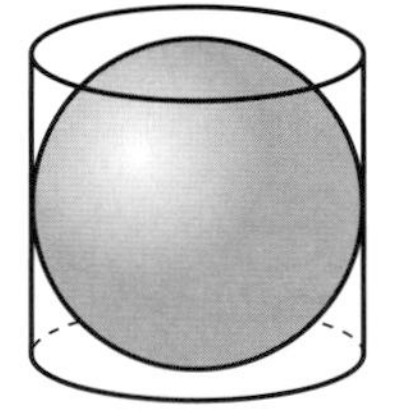

7 Calculate the volume of a pyramid with a square base of side 5.3 cm and a height of 6.3 cm.

8 A sector of a circle has a radius of 7.2 cm and an area of 50 cm^2.

Calculate the angle θ.

θ

A*

9 A frustum of a cone has a base of radius 12 cm, and a top of radius 3 cm.

The height of the frustum is 6 cm.

Calculate the volume of the frustum.

Geometry and measure

2 Angles and polygons

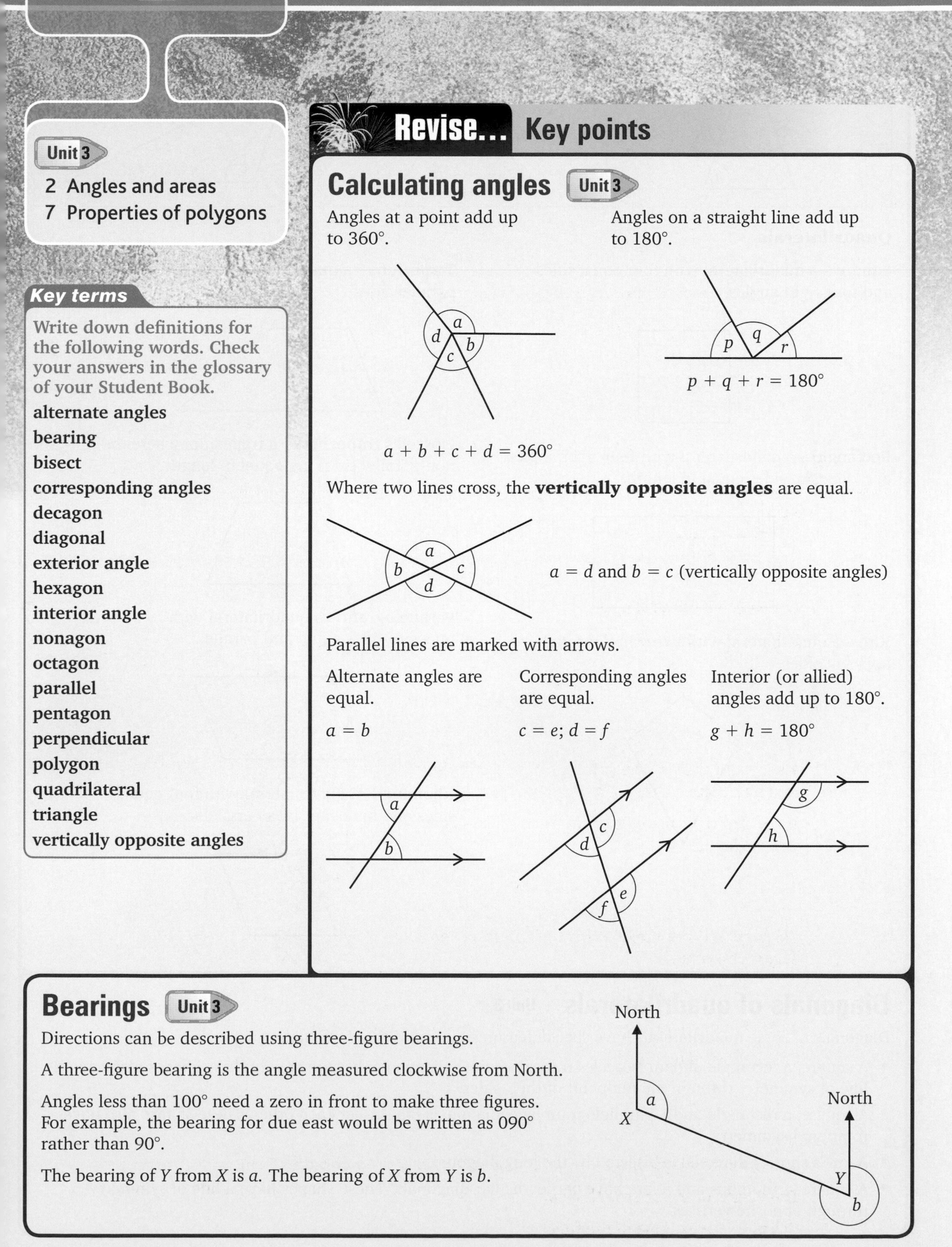

Unit 3

2 Angles and areas
7 Properties of polygons

Key terms

Write down definitions for the following words. Check your answers in the glossary of your Student Book.

alternate angles
bearing
bisect
corresponding angles
decagon
diagonal
exterior angle
hexagon
interior angle
nonagon
octagon
parallel
pentagon
perpendicular
polygon
quadrilateral
triangle
vertically opposite angles

Revise... Key points

Calculating angles (Unit 3)

Angles at a point add up to 360°.

$a + b + c + d = 360°$

Angles on a straight line add up to 180°.

$p + q + r = 180°$

Where two lines cross, the **vertically opposite angles** are equal.

$a = d$ and $b = c$ (vertically opposite angles)

Parallel lines are marked with arrows.

Alternate angles are equal.

$a = b$

Corresponding angles are equal.

$c = e$; $d = f$

Interior (or allied) angles add up to 180°.

$g + h = 180°$

Bearings (Unit 3)

Directions can be described using three-figure bearings.

A three-figure bearing is the angle measured clockwise from North.

Angles less than 100° need a zero in front to make three figures. For example, the bearing for due east would be written as 090° rather than 90°.

The bearing of Y from X is a. The bearing of X from Y is b.

Angles and shapes Unit 3

The angles in a triangle add up to 180°.

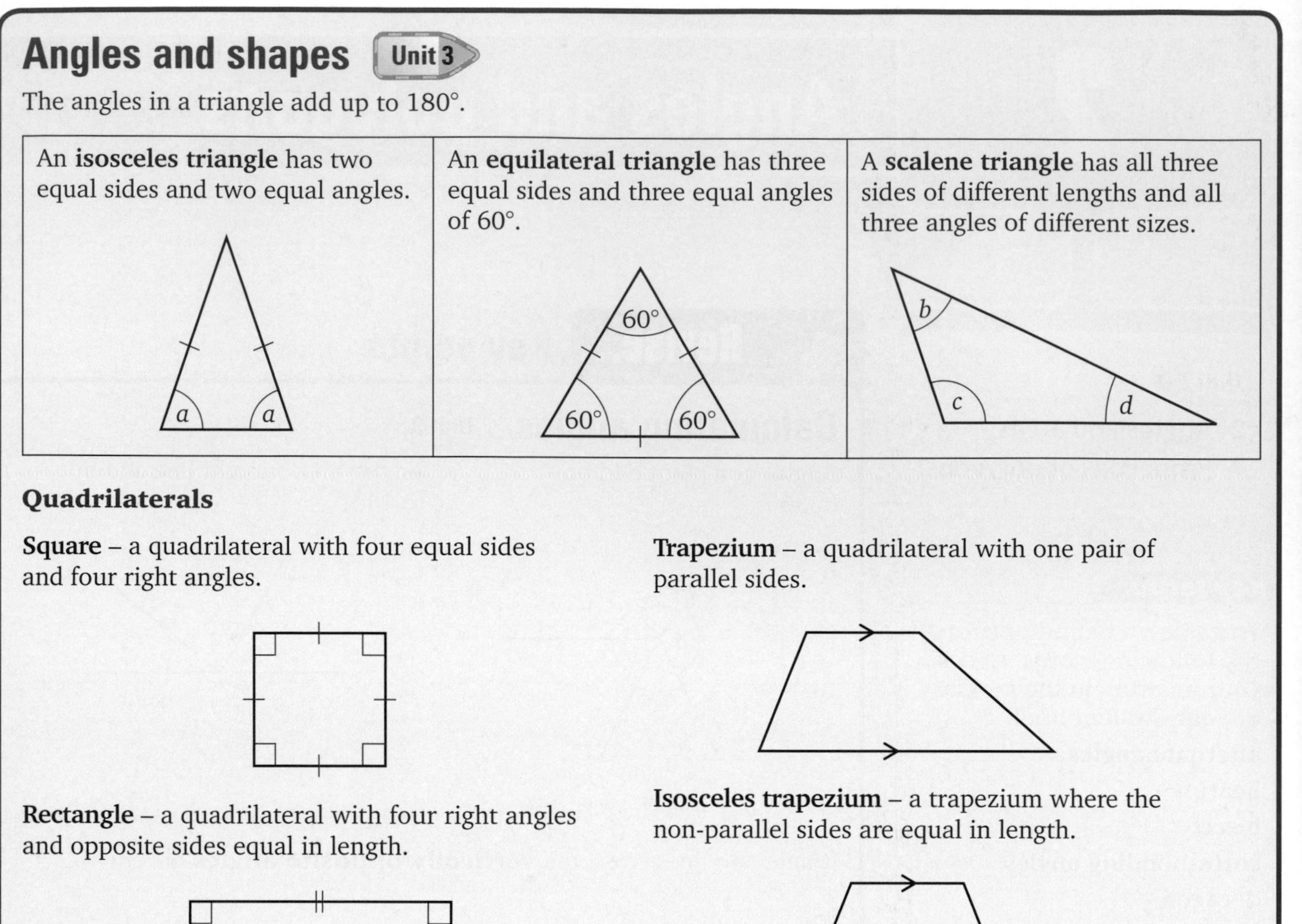

An **isosceles triangle** has two equal sides and two equal angles.	An **equilateral triangle** has three equal sides and three equal angles of 60°.	A **scalene triangle** has all three sides of different lengths and all three angles of different sizes.

Quadrilaterals

Square – a quadrilateral with four equal sides and four right angles.

Rectangle – a quadrilateral with four right angles and opposite sides equal in length.

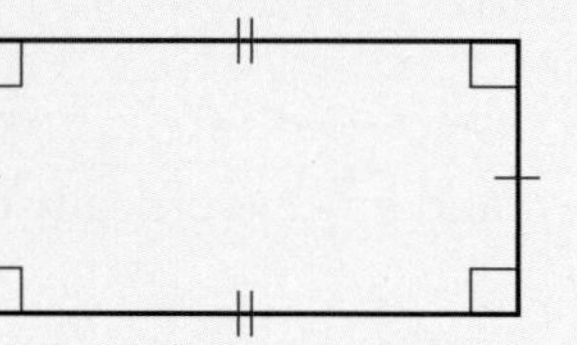

Kite – a quadrilateral with two pairs of equal adjacent sides.

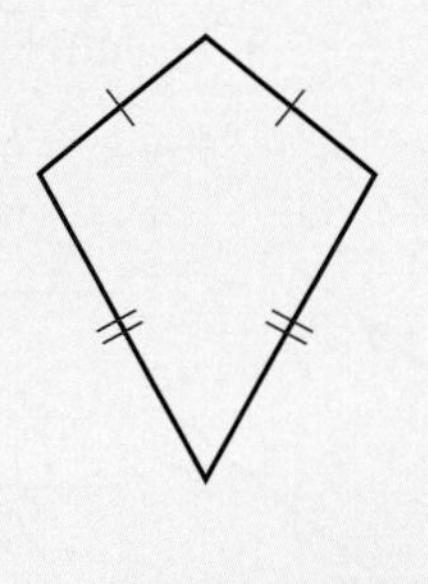

Trapezium – a quadrilateral with one pair of parallel sides.

Isosceles trapezium – a trapezium where the non-parallel sides are equal in length.

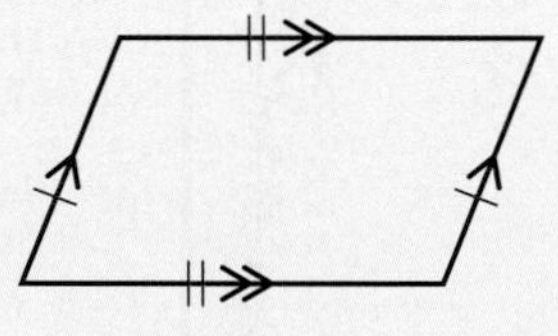

Parallelogram – a quadrilateral with opposite sides equal and parallel.

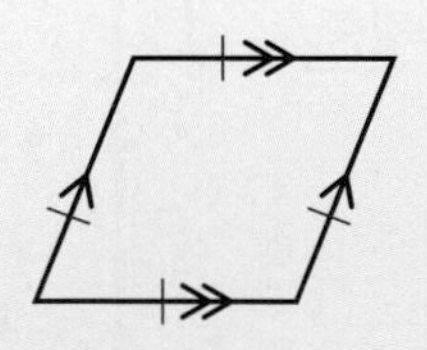

Rhombus – a quadrilateral with four equal sides and opposite sides parallel.

Diagonals of quadrilaterals Unit 3

Diagonals of some quadrilaterals have special features, linked to their symmetry properties.

- A square, a rectangle and an isosceles trapezium have diagonals that are equal. These shapes also have a line of symmetry through the midpoint of their sides.
- A square, a rectangle and a parallelogram have diagonals that bisect each other. These shapes also have rotational symmetry.
- A kite's shorter diagonal is bisected by the long diagonal.
- A square, a rhombus and a kite have perpendicular diagonals. These shapes have a line of symmetry through opposite vertices.

Angles in polygons Unit 3

A polygon can always be divided into triangles. The number of triangles is always 2 less than the number of sides in the polygon.

The sum of the angles is always (number of sides − 2) × 180°.

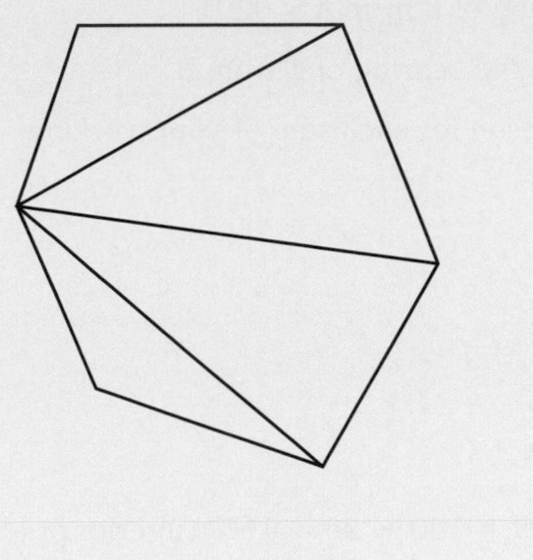

Bump up your grade

Learn the rule to work out the sum of angles in any polygon.

In a regular polygon, all angles are equal.

In all polygons, the exterior angles add up to 360°.

AQA Examiner's tip

Remember that in a regular polygon, all interior angles are equal and all exterior angles are equal.

Example Naming and calculating angles Unit 3

D

In the diagram, *AB* and *CD* are parallel.

Angle *AEF* = 42° and *DGH* = 53°.

a Calculate angles:

i *EFG*

ii *FEG*, giving reasons for your answers.

b What type of angle is angle *CFE*?

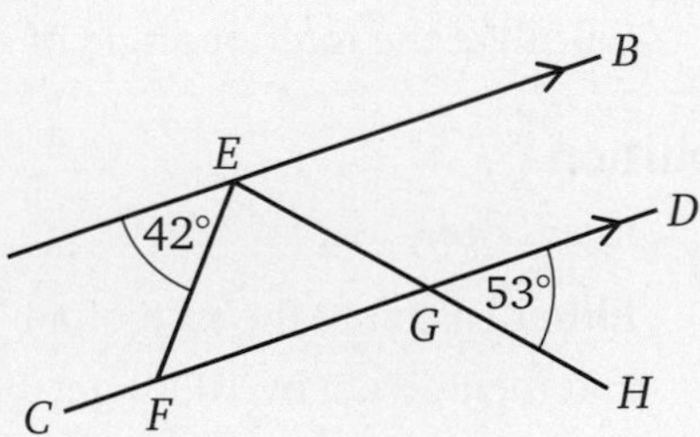

AQA Examiner's tip

When you work out an angle, write it on the diagram, as you may need it later.

Solution

a **i** *EFG* = *AEF* = 42° (alternate angles on parallel lines)

ii *BEG* = *DGH* = 53° (corresponding angles on parallel lines)

FEG = 180 − 42 − 53 = 85° (angles on a straight line = 180°)

b *CFE* is obtuse.

AQA Examiner's tip

Use the correct terms to describe the angles. For example, call them alternate angles, not Z angles.

Example Bearings Unit 3

C

A ship sails in an equilateral triangle from A to B to C.

The bearing of B from A is 020°.

Calculate the bearing of C from B.

Give a reason for each step of your working.

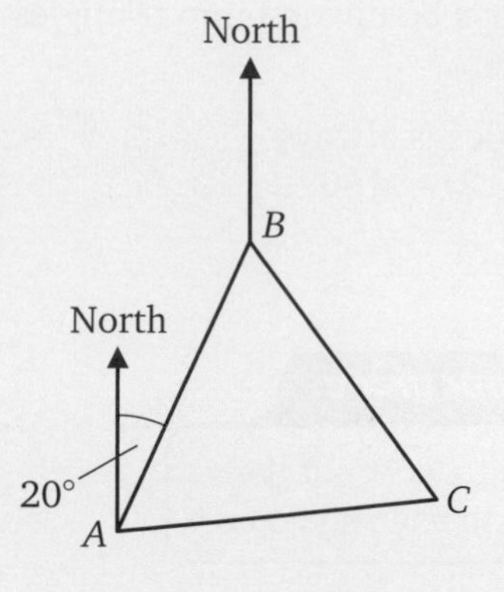

Solution

Angle $ABN = 180 - 20 = 160°$ (interior or allied angles)

Angle $ABC = 60°$ (equilateral triangle)

$NBC = 360 - 60 - 160$ (angles round a point = 360°)

$NBC = 140°$

This is the bearing of C from B.

Example Polygons Unit 3

C

a A quadrilateral has two pairs of equal sides. Its diagonals are perpendicular but not equal. What type of quadrilateral is it?

b Calculate the interior angle of a regular decagon.

Solution

a It is a kite.

b **Either** calculate the sum of all 10 angles: $(10 - 2) \times 180 = 1440$
and then divide by 10, to get $1440 \div 10 = 144$
or divide 360 by 10 to get each exterior angle: $360 \div 10 = 36$
and then subtract from 180 to get the interior angle, $180 - 36 = 144°$

Practise... Angles and polygons Unit 3

D C B

D

1 Measure the bearing of:

a B from A

b C from B

c A from C.

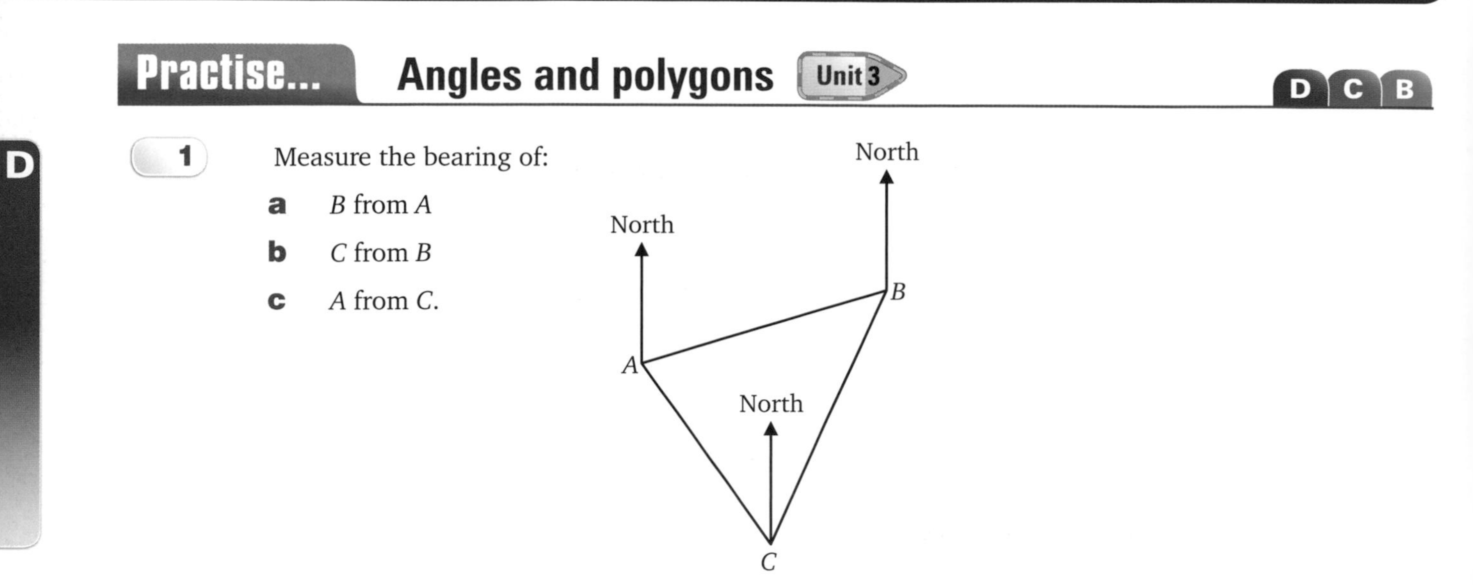

2 *AB* and *CD* are parallel.

Find the value of:

a x

b y

c z, giving a reason for each answer.

Not drawn accurately

D

3 The diagram shows an equilateral triangle, a regular hexagon and a regular octagon.

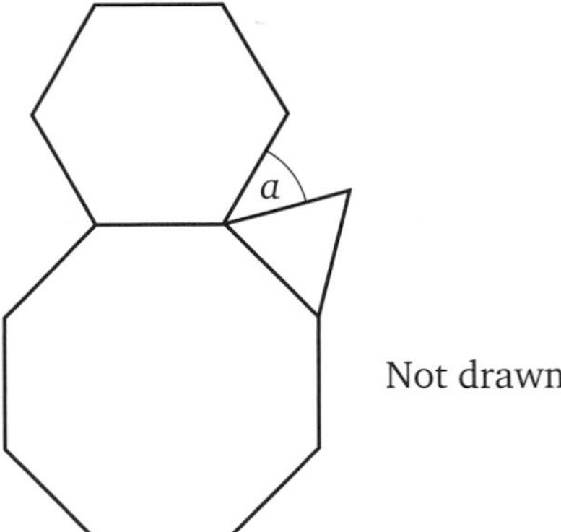

Not drawn accurately

Calculate the size of angle a.

C

4 A regular polygon has exterior angles of 24°.

How many sides does it have?

5 Explain why regular hexagons will tessellate but regular pentagons do not.

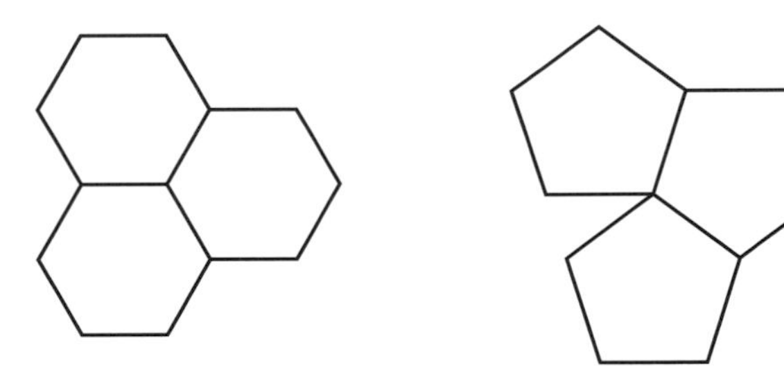

6 The diagram shows a regular pentagon *ABCDE* and a diagonal *AD*.

a Calculate the size of angles, giving a reason for your answers.

i *BCD*

ii *EDA*

iii *ADC*

b How do your answers show that *AD* and *BC* are parallel?

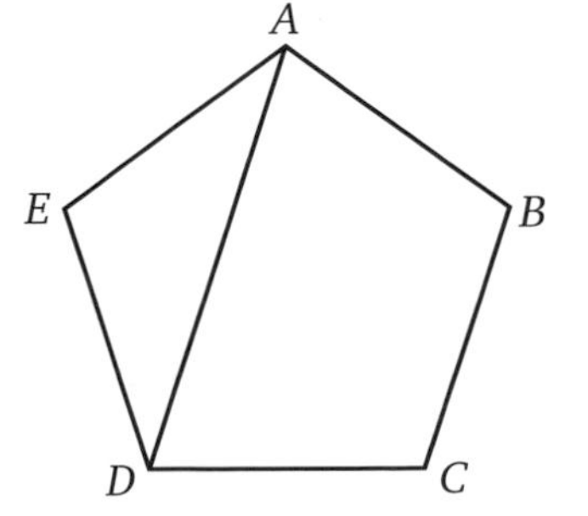

7 A pentagon has two angles each of size $x°$, one angle of size $2x°$, one of size $x + 10°$, and one of size 100°.

Calculate the value of x.

B

8 Steve wanted to draw a pentagon with 4 equal angles, and the fifth angle twice the size of each of the other four.

Explain why this is impossible.

Geometry and measure

3 Circle theorems

Unit 3

Key terms

Write down definitions for the following words. Check your answers in the glossary of your Student Book.

alternate segment
angle subtended
arc (of a circle)
chord
cyclic quadrilateral
supplementary angles
tangent

Revise... Key points

Circle theorems Unit 3

There are four angle properties of circles you need to know.

1 **The angle subtended by an arc at the centre of a circle is twice the angle subtended at the circumference.**

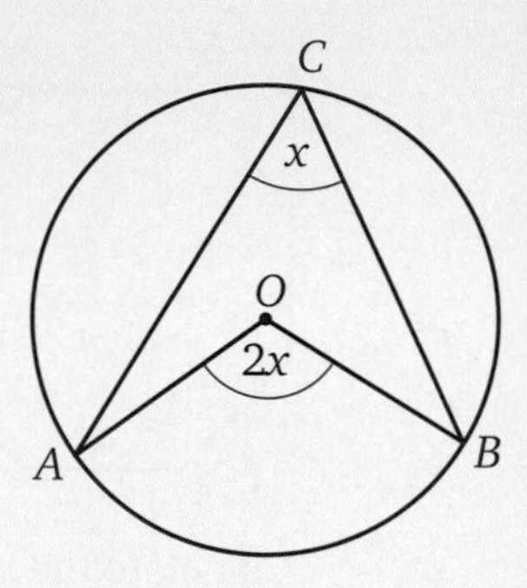

2 **The angle in a semicircle is a right angle.**

O is the centre of the circle, so the line through O is a diameter.

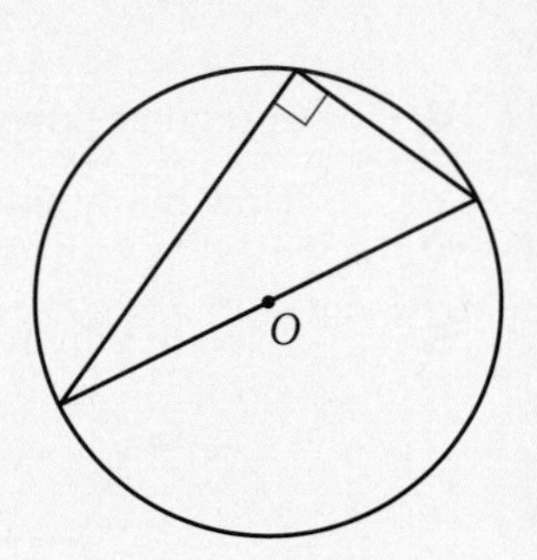

3 **Angles subtended by the same arc are equal.**

$a = b$ as they are both subtended by the same arc.

$c = d$ as they are both subtended by the same arc.

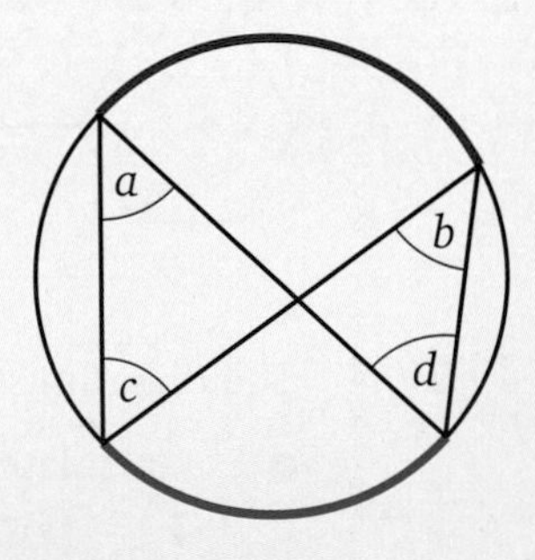

4 **The opposite angles of a cyclic quadrilateral add up to 180°.**

A cyclic quadrilateral is a four-sided shape where the four vertices touch a circle.

$a + d = 180°$, $b + c = 180°$

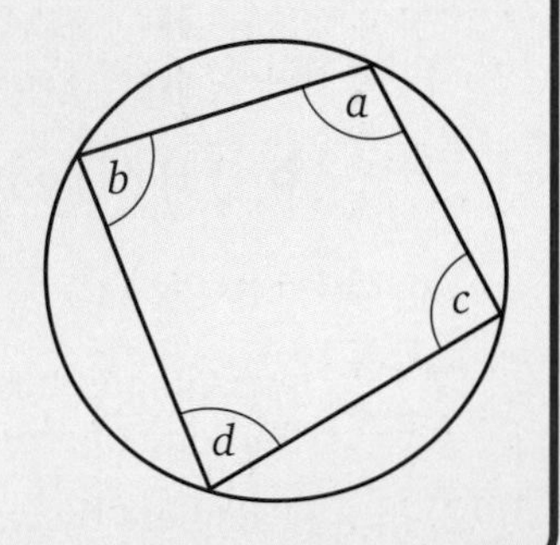

Tangents and chords Unit 3

Tangent

A **tangent** is a straight line which touches a circle at one point only.

PQ is a tangent to the circle. It touches the circle at *A*.

Chord

A **chord** is a straight line joining two points on the circumference of a circle. *BC* is a chord of the circle.

A diameter is a chord which passes through the centre of the circle.

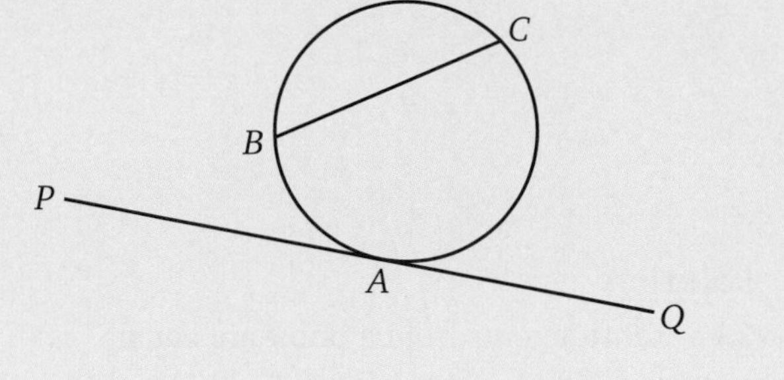

Tangent properties

You need to know the following tangent properties:

1. **Tangents from the same point to the circle are equal.**
   ($PM = PN$)
2. **The angle between a tangent and a radius = 90°.**
   ($PMO = PNO = 90°$)

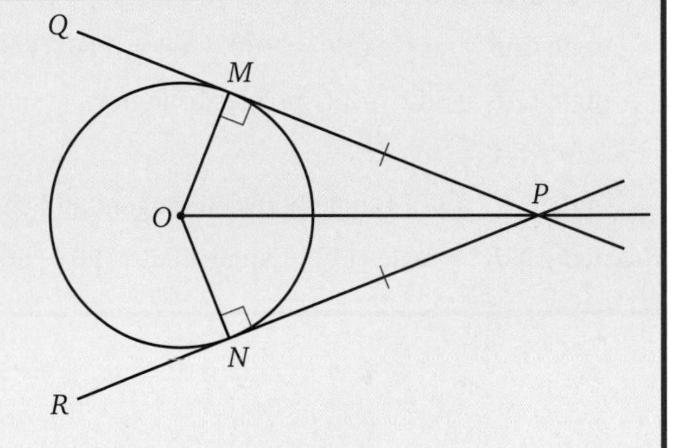

Chord properties

You need to know the following chord properties:

1. **The perpendicular line from the centre of the circle to a chord bisects the chord.**
   (If angle $ODA = 90°$, then $AD = DB$)
2. **The perpendicular bisector of any chord passes through the centre of the circle.**
   (If $AD = DB$ and $ODA = 90°$, then *O* is the centre of the circle.)

Alternate segment theorem

In the diagram, the chord *AC* divides the circle into two segments.

The chord makes an angle *CAP* with the tangent.

The angle *ABC* is called the angle in the **alternate segment**, as it is on the other side of the chord *AC*.

The alternate segment theorem says that the angle between the tangent and the chord is the same as the angle in the alternate segment.

Angle *PAC* = angle *ABC*

The chord AB also splits the circle into two segments, so angle *BAQ* = angle *BCA*

AQA Examiner's tips

Students often make the mistake of saying angle $PAC = ACB$ (alternate angles), but *PQ* and *CB* are not parallel. Make sure you understand the alternate segment theorem.

Example Tangents and chords Unit 3

B

In the diagram, AC and BC are tangents to the circle with centre O.

Angle $ABC = 68°$

Calculate angle OAB.

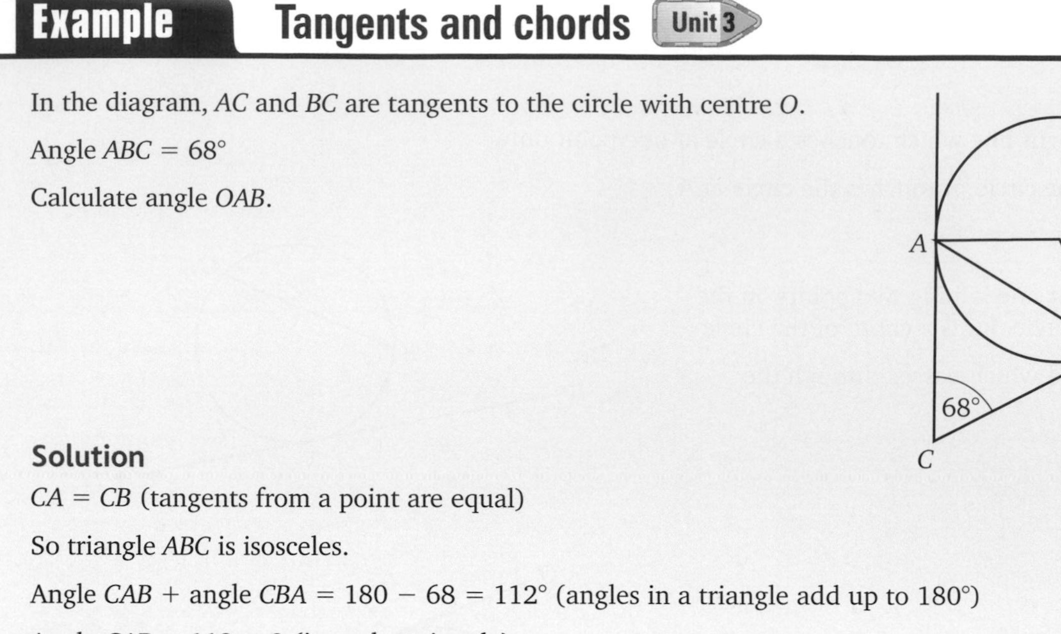

Solution

$CA = CB$ (tangents from a point are equal)

So triangle ABC is isosceles.

Angle CAB + angle $CBA = 180 - 68 = 112°$ (angles in a triangle add up to 180°)

Angle $CAB = 112 \div 2$ (isosceles triangle)

$CAB = 56°$

Angle $OAC = 90°$ (angle between tangent and chord = 90°)

Angle OAB = angle OAC − angle $CAB = 90 - 56 = 34°$

Example Circle theorems Unit 3

A

A, B and C are points on the circumference of a circle, centre O.

AD is a tangent to the circle.

$ABC = 50°$ and $BAD = 55°$

Calculate angles ACB and CBO, giving reasons for each step.

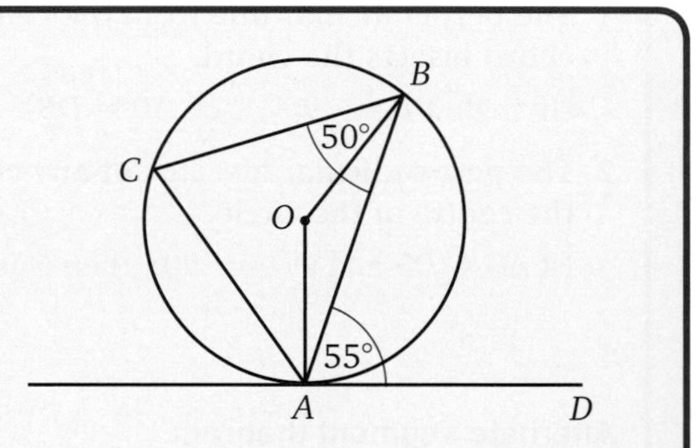

Solution

$ACB = BAD = 55°$ (alternate segment)

$AOB = 2 \times ACB = 110°$ (angle at centre twice angle at circumference)

$AO = OB$ (radii)

$OBA = (180 - 110) \div 2 = 35°$ (base angle, isosceles triangle)

$CBO = CBA - OBA = 50 - 35 = 15°$

AQA Examiner's tip

You will not get full marks if you don't show your method.
You should use the correct vocabulary.

Practise... Circle theorems Unit 3

B A A*

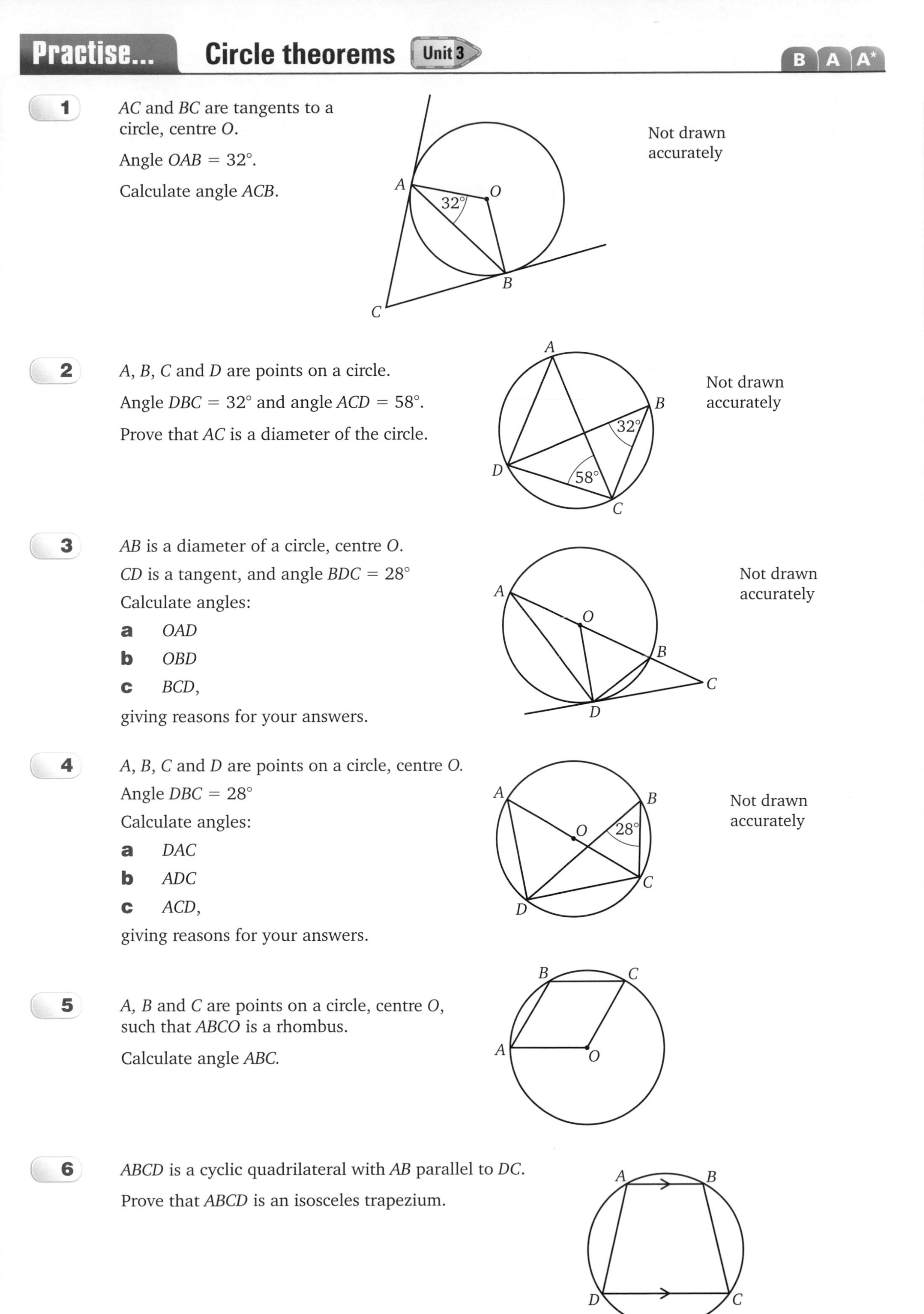

1 *AC* and *BC* are tangents to a circle, centre *O*.

Angle *OAB* = 32°.

Calculate angle *ACB*.

2 *A*, *B*, *C* and *D* are points on a circle.

Angle *DBC* = 32° and angle *ACD* = 58°.

Prove that *AC* is a diameter of the circle.

3 *AB* is a diameter of a circle, centre *O*.

CD is a tangent, and angle *BDC* = 28°

Calculate angles:

a *OAD*

b *OBD*

c *BCD*,

giving reasons for your answers.

4 *A*, *B*, *C* and *D* are points on a circle, centre *O*.

Angle *DBC* = 28°

Calculate angles:

a *DAC*

b *ADC*

c *ACD*,

giving reasons for your answers.

5 *A*, *B* and *C* are points on a circle, centre *O*, such that *ABCO* is a rhombus.

Calculate angle *ABC*.

6 *ABCD* is a cyclic quadrilateral with *AB* parallel to *DC*.

Prove that *ABCD* is an isosceles trapezium.

Geometry and measure

4 Transformations and vectors

Unit 3

9 Rotations, reflections and translations
14 Enlargements
19 Vectors

Key terms

Write down definitions for the following words. Check your answers in the glossary of your Student Book.

angle of rotation
centre of enlargement
centre of rotation
coordinate
enlargement
line of symmetry
reflection
rotation
scale factor
similar
congruent
transformation
translation
vector
vertex (vertices)

Revise... Key points

Reflection Unit 3

In all **transformations**, the original position is called the **object**, and the result of the transformation is called the **image**.

Reflection symmetry is a transformation where half a shape is a mirror image of the other half. The **mirror line** is called the **line of symmetry**.

Mirror lines are usually defined by equations.

Here, the triangle has been reflected in the line $x = 2$.

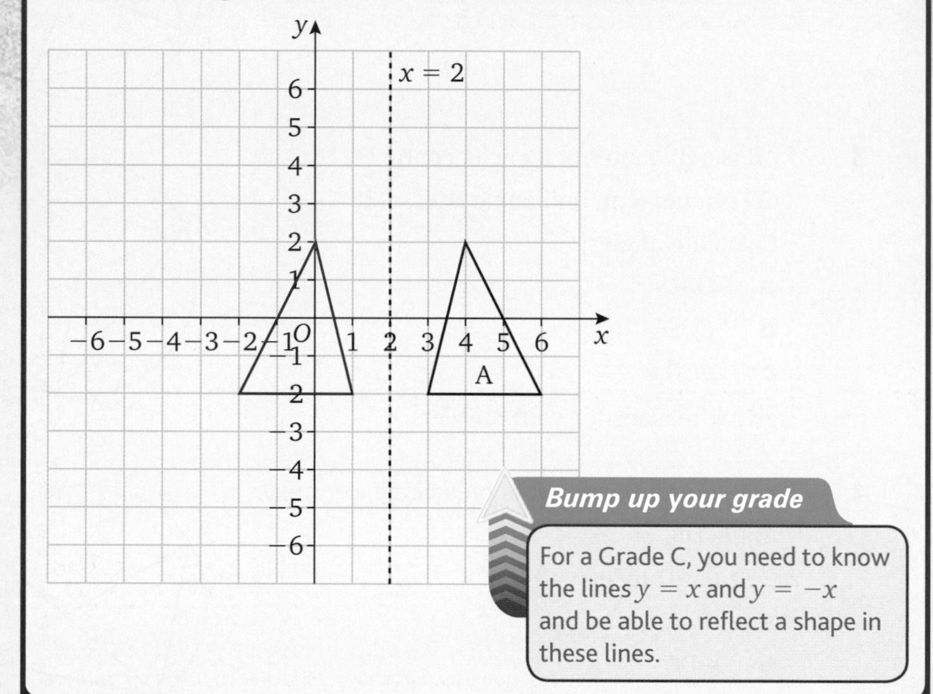

Bump up your grade

For a Grade C, you need to know the lines $y = x$ and $y = -x$ and be able to reflect a shape in these lines.

Rotation Unit 3

To specify a **rotation**, you need to give the **centre of rotation**, and the **angle of rotation** and direction (clockwise or anticlockwise).

You should use tracing paper to help with rotations.

In the diagram, shape A has been rotated 90° anticlockwise.

The centre of rotation is $(1, -2)$.

Imagine a wheel with its centre at $(1, -2)$.

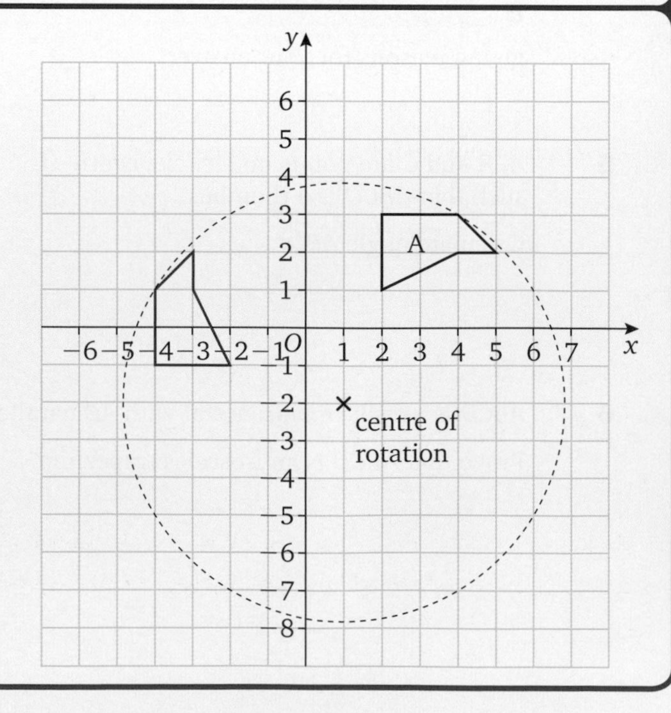

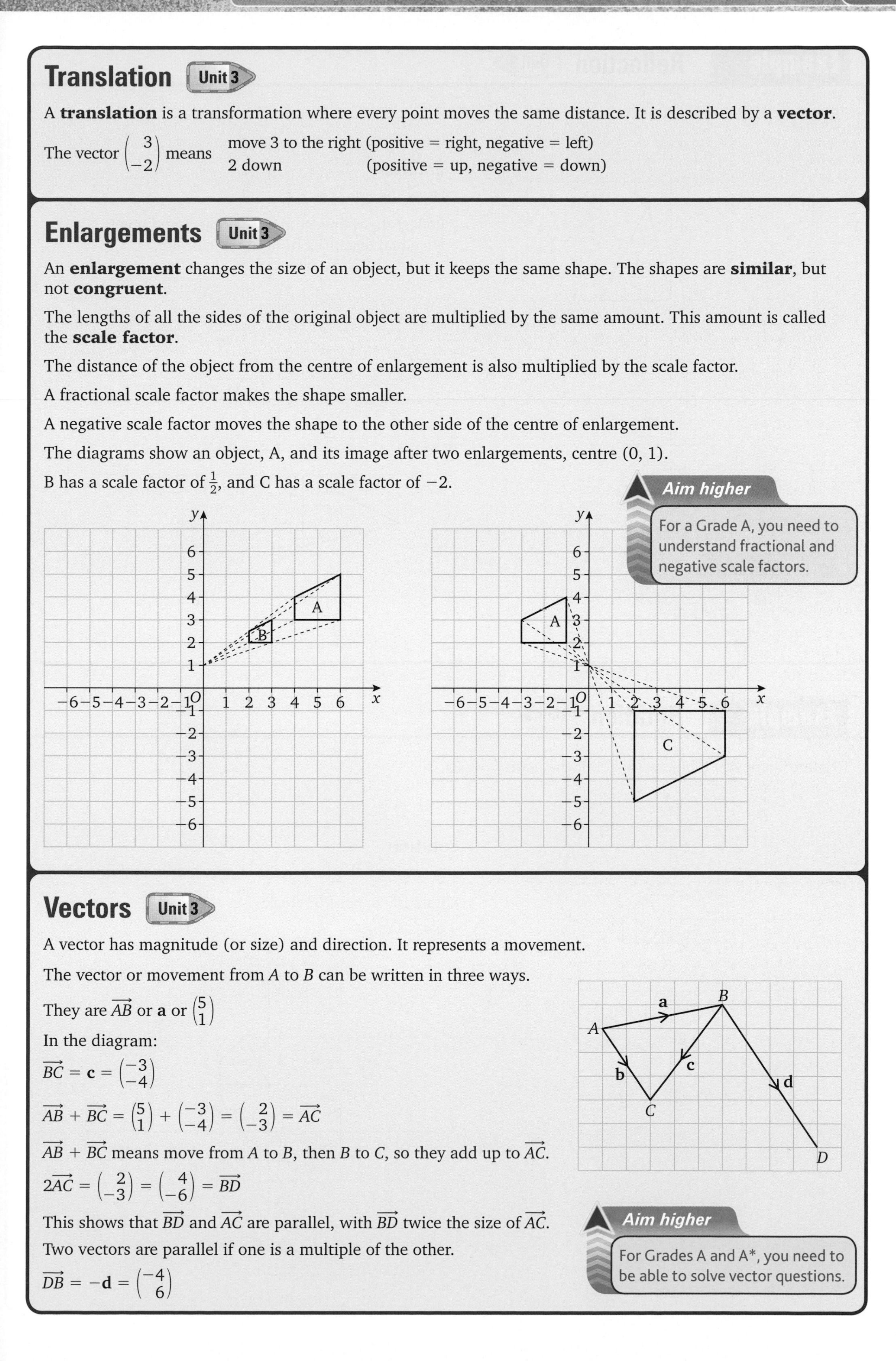

Translation Unit 3

A **translation** is a transformation where every point moves the same distance. It is described by a **vector**.

The vector $\begin{pmatrix} 3 \\ -2 \end{pmatrix}$ means move 3 to the right (positive = right, negative = left)
2 down (positive = up, negative = down)

Enlargements Unit 3

An **enlargement** changes the size of an object, but it keeps the same shape. The shapes are **similar**, but not **congruent**.

The lengths of all the sides of the original object are multiplied by the same amount. This amount is called the **scale factor**.

The distance of the object from the centre of enlargement is also multiplied by the scale factor.

A fractional scale factor makes the shape smaller.

A negative scale factor moves the shape to the other side of the centre of enlargement.

The diagrams show an object, A, and its image after two enlargements, centre (0, 1).

B has a scale factor of $\frac{1}{2}$, and C has a scale factor of -2.

Aim higher

For a Grade A, you need to understand fractional and negative scale factors.

Vectors Unit 3

A vector has magnitude (or size) and direction. It represents a movement.

The vector or movement from A to B can be written in three ways.

They are $\overrightarrow{AB}$ or **a** or $\begin{pmatrix} 5 \\ 1 \end{pmatrix}$

In the diagram:

$\overrightarrow{BC} = \mathbf{c} = \begin{pmatrix} -3 \\ -4 \end{pmatrix}$

$\overrightarrow{AB} + \overrightarrow{BC} = \begin{pmatrix} 5 \\ 1 \end{pmatrix} + \begin{pmatrix} -3 \\ -4 \end{pmatrix} = \begin{pmatrix} 2 \\ -3 \end{pmatrix} = \overrightarrow{AC}$

$\overrightarrow{AB} + \overrightarrow{BC}$ means move from A to B, then B to C, so they add up to $\overrightarrow{AC}$.

$2\overrightarrow{AC} = \begin{pmatrix} 2 \\ -3 \end{pmatrix} = \begin{pmatrix} 4 \\ -6 \end{pmatrix} = \overrightarrow{BD}$

This shows that $\overrightarrow{BD}$ and $\overrightarrow{AC}$ are parallel, with $\overrightarrow{BD}$ twice the size of $\overrightarrow{AC}$.

Two vectors are parallel if one is a multiple of the other.

$\overrightarrow{DB} = -\mathbf{d} = \begin{pmatrix} -4 \\ 6 \end{pmatrix}$

Aim higher

For Grades A and A*, you need to be able to solve vector questions.

Example Reflection Unit 3

C

Reflect triangle A in the line $y = -x$

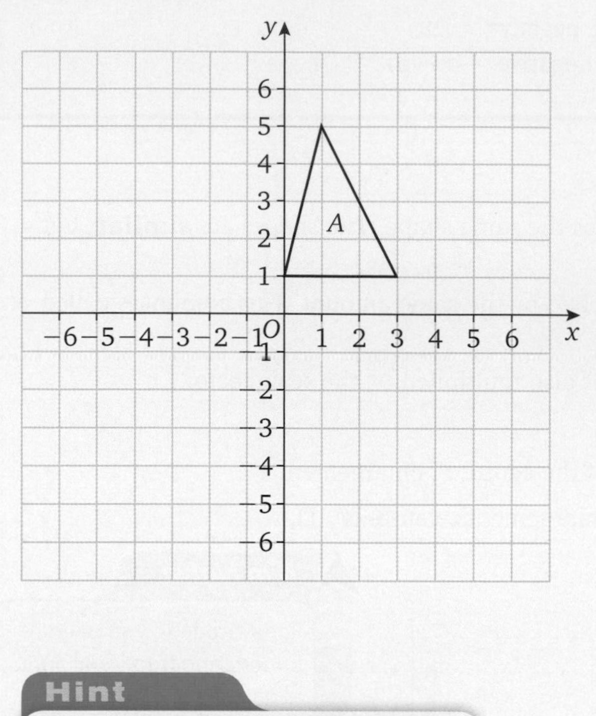

Solution

Draw the line $y = -x$

Reflect the shape, so that the object and image are equal distances from the mirror line.

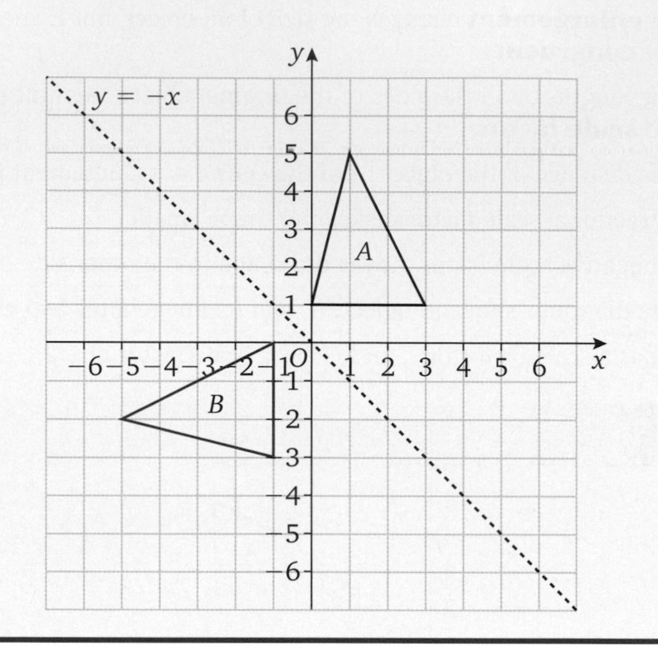

Hint

It can help to rotate the page so the mirror line is vertical.

Example Rotation Unit 3

C

Rotate shape A 90° clockwise about the point $(-2, 3)$.

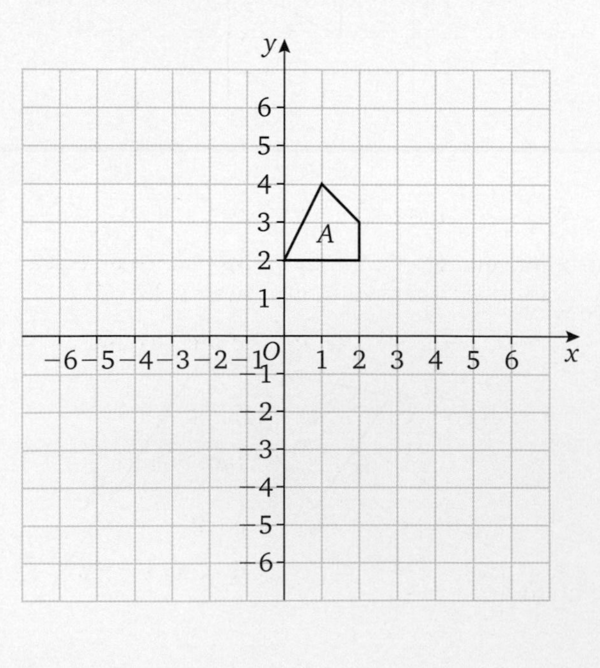

Solution

Use tracing paper; trace the shape, hold it still at $(-2, 3)$.

Rotate the paper 90° clockwise.

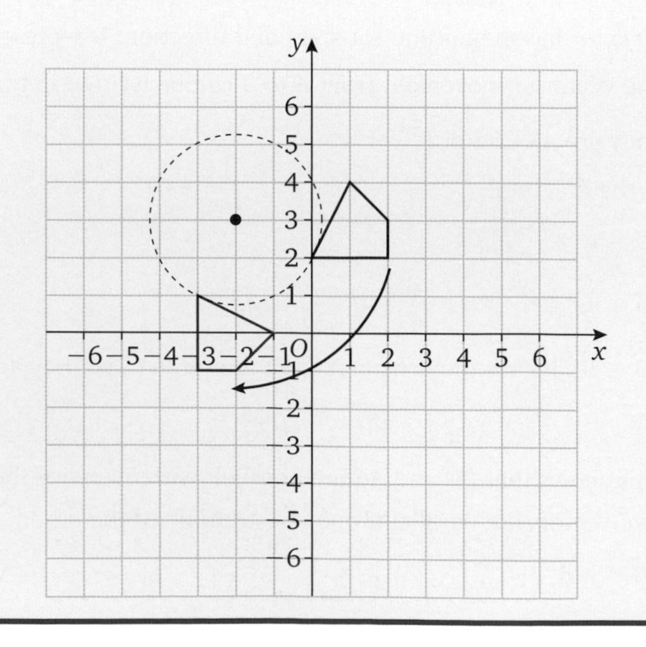

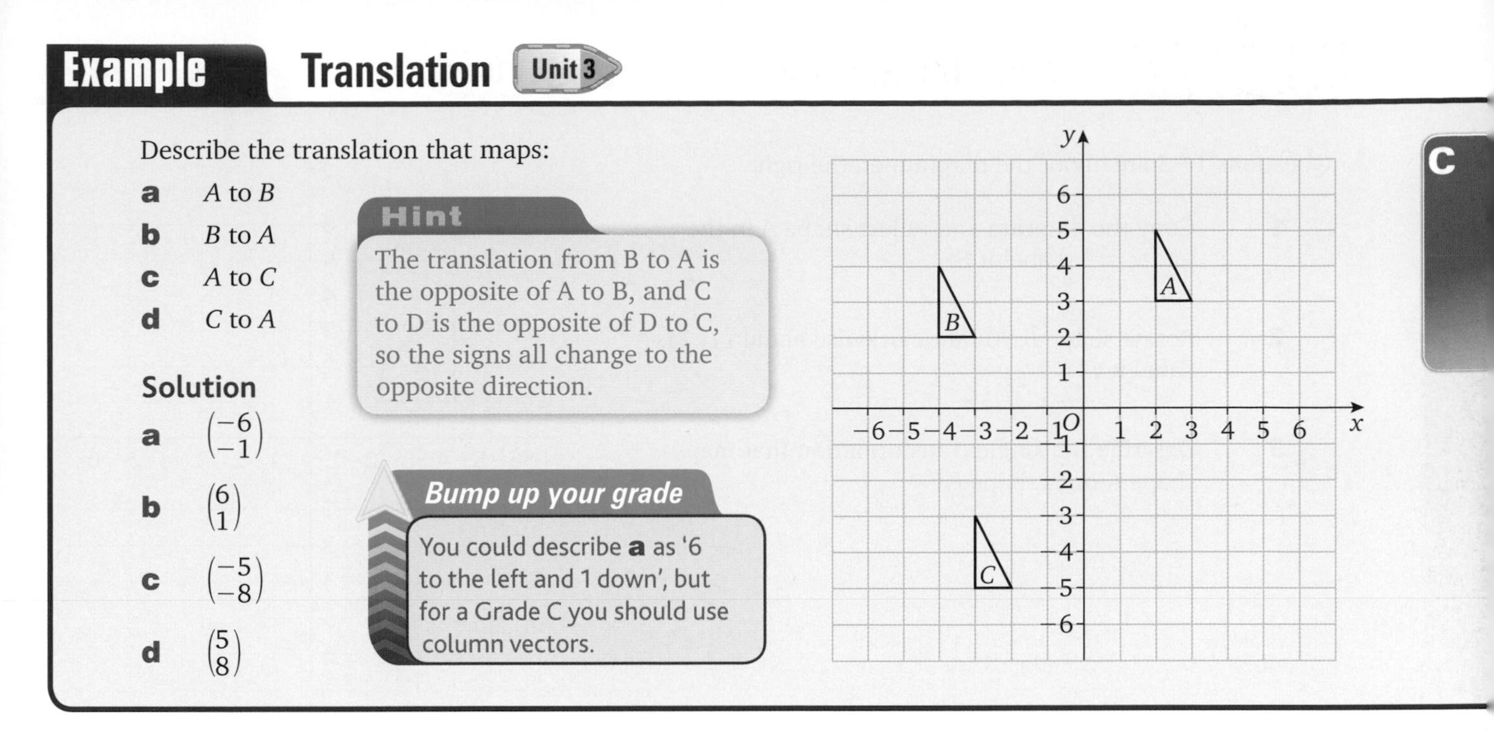

Example Translation Unit 3

Describe the translation that maps:

a A to B

b B to A

c A to C

d C to A

Hint

The translation from B to A is the opposite of A to B, and C to D is the opposite of D to C, so the signs all change to the opposite direction.

Solution

a $\begin{pmatrix} -6 \\ -1 \end{pmatrix}$

b $\begin{pmatrix} 6 \\ 1 \end{pmatrix}$

c $\begin{pmatrix} -5 \\ -8 \end{pmatrix}$

d $\begin{pmatrix} 5 \\ 8 \end{pmatrix}$

Bump up your grade

You could describe **a** as '6 to the left and 1 down', but for a Grade C you should use column vectors.

C

Example Enlargements Unit 3

a Enlarge the trapezium with a scale factor of 1.5, centre $(-5, 4)$.

b Calculate the areas of the object and the image.

C

Solution

a

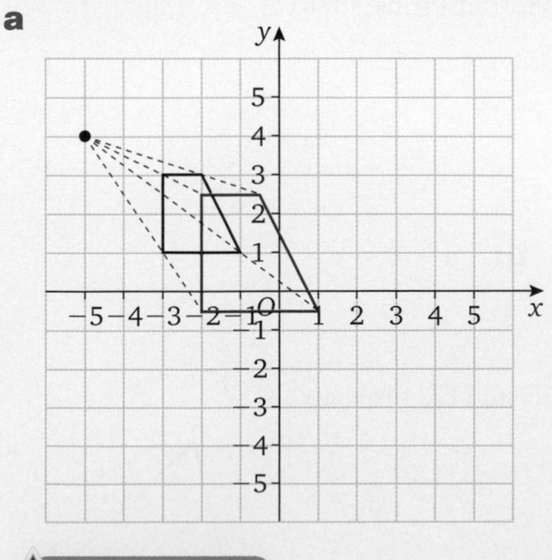

Mark the centre of enlargement.

Draw lines from the centre through each vertex and extend them to three times the length.

Join the ends of these lines to produce the enlarged shape.

b The object is a trapezium of area $\frac{1+2}{2} \times 2 = 3$ square units.

The image is a trapezium of area $\frac{1.5+3}{2} \times 3 = 6.75$ square units.

Aim higher

For a Grade A, you need to know that, after an enlargement with scale factor s, the area is increased by a scale factor s^2, and the volume is increased by a scale factor s^3.
In the example, the scale factor was 1.5, and the area is increased by $1.5^2 = 2.25$, as $3 \times 2.25 = 6.75$

Practise... Transformations and vectors

Unit 3

D C B A A*

Questions 1–3 are about the diagram on the right.

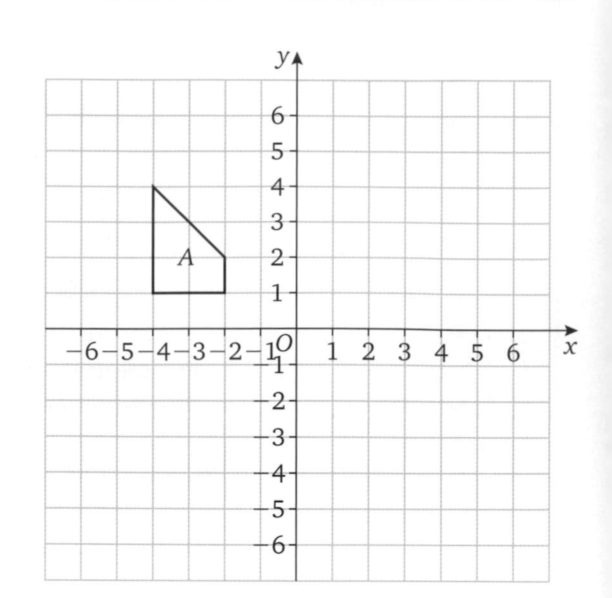

D **1** Copy the diagram and reflect shape A in the line $x = 1$. Label it B.

C **2** Rotate shape B 90° anticlockwise about (1, 1). Label it C.

3 Describe the single transformation that maps shape A onto shape C.

Questions 4–7 are about the diagram on the right.

Copy the diagram for each question.

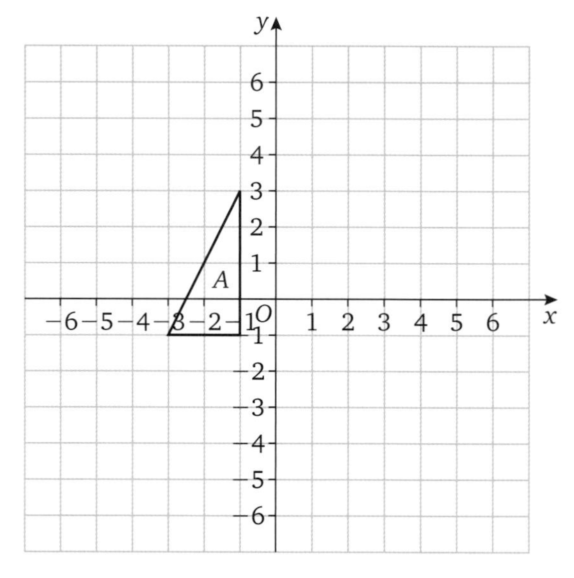

C **4** Translate triangle A through $\begin{pmatrix} 4 \\ -2 \end{pmatrix}$.

5 Enlarge triangle A by a scale factor of $\frac{1}{2}$, centre (5, 1).

A **6** Enlarge triangle A by a scale factor of −2, centre (0, 1).

B **7** **a** Reflect triangle A in the line $x = 1$. Label the image B.

b Reflect B in the line $y = -2$. Label the image C.

c Describe fully the single transformation that transforms A into C.

A **8** $\mathbf{a} = \begin{pmatrix} 3 \\ -1 \end{pmatrix}$, $\mathbf{b} = \begin{pmatrix} 1 \\ -2 \end{pmatrix}$, $\mathbf{c} = \begin{pmatrix} 4 \\ 2 \end{pmatrix}$

a Calculate:

i $\mathbf{a} + \mathbf{b}$ **ii** $2\mathbf{b} - \mathbf{a}$ **iii** $\mathbf{a} + \mathbf{b} - \mathbf{c}$

b Show that $\mathbf{c}$ is parallel to $\mathbf{a} - \mathbf{b}$.

A*  In triangle FGH, M and N are the midpoints of FH and GH respectively.

$\overrightarrow{FM} = \mathbf{a}$ and $\overrightarrow{GN} = \mathbf{b}$

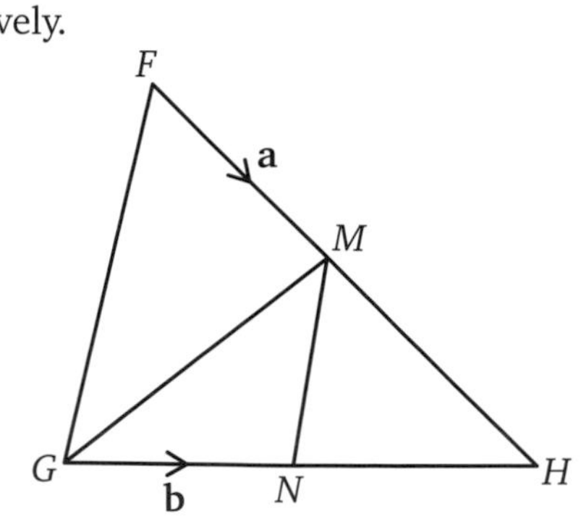

a Find, in terms of **a** and **b**,

i $\overrightarrow{GH}$

ii $\overrightarrow{MN}$

iii $\overrightarrow{GM}$

iv $\overrightarrow{FG}$

b What can you say about FG and MN?

Geometry and measure

5 Measures, loci and construction

Unit 3

8 Coordinates and graphs
12 Measures
15 Constructions
16 Loci

Key terms

Write down definitions for the following words. Check your answers in the glossary of your Student Book.

- arc (of a circle)
- bisect, bisector
- compound measure
- congruent
- elevation
- equidistant
- equilateral
- face
- locus, loci
- mass
- net
- perpendicular
- plan
- scale
- similar
- vertex

Revise... Key points

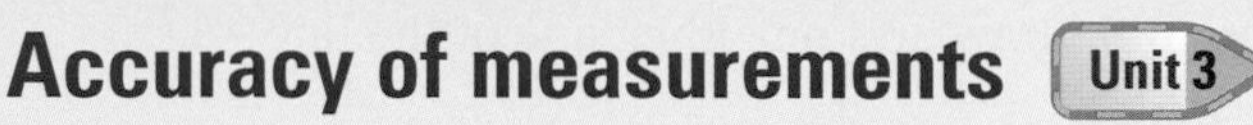

Accuracy of measurements (Unit 3)

A line, given as 6 cm long to the nearest centimetre, could be as small as 5.5 cm. This is the lower bound.

The upper bound (the largest possible value) is given as 6.5 cm. In reality, 6.5 cm would be rounded up to 7 cm, but the largest possible value is $6.4\dot{9}$, which is essentially the same as 6.5 cm.

'To the nearest centimetre' means the actual distance could be any value from half a centimetre below to half a centimetre above.

Compound measures (Unit 3)

Compound measures combine two units.

Use the units to tell you what calculation to do.

To calculate a speed in km/h, remember km/h means $\frac{\text{km}}{\text{h}}$ or km ÷ hours.

A speed in km/h is how many km you would travel if you continue at the same rate for 1 hour.

Fuel consumption in km/l means $\frac{\text{km}}{\text{l}}$ or km ÷ litres used.

This is the number of km travelled on 1 litre of fuel.

Density in g/cm^3 means g ÷ cm^3.

This is the **mass** of 1 cm^3 of a substance.

Constructing triangles (Unit 3)

Given a side and 2 angles

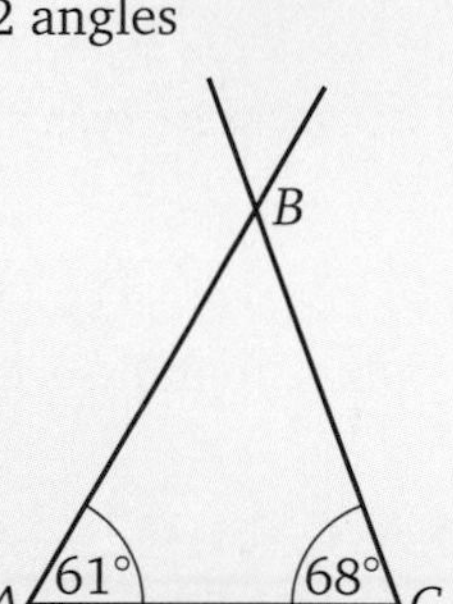

Draw *AC* 7 cm long.
Measure angle *BAC* = 61°.
Draw a long line.
Measure angle *BCA* = 68°.
Draw a long line to cross at *B*.

Given 2 sides and an angle

B
8 cm
A
61°
C
7 cm

Draw *AC* 7 cm long.
Measure angle *BAC* = 61°.
Measure *AB* = 8 cm.
Join *B* to *C*.

Given 3 sides

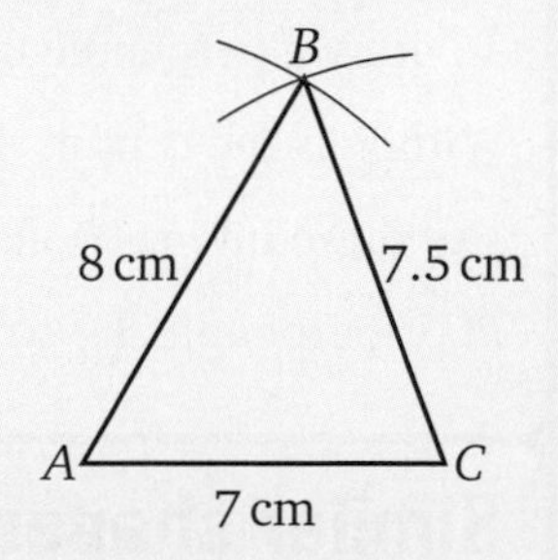

Draw *AC* 7 cm long.
Open compasses to 8 cm.
With compass point on *A*, draw an **arc**.
Open compasses to 7.5 cm.
With compass point on *C*, draw an arc. The arcs intersect at *B*.

AQA Examiner's tip

Remember to leave your arcs as they are your working out, and show your method.

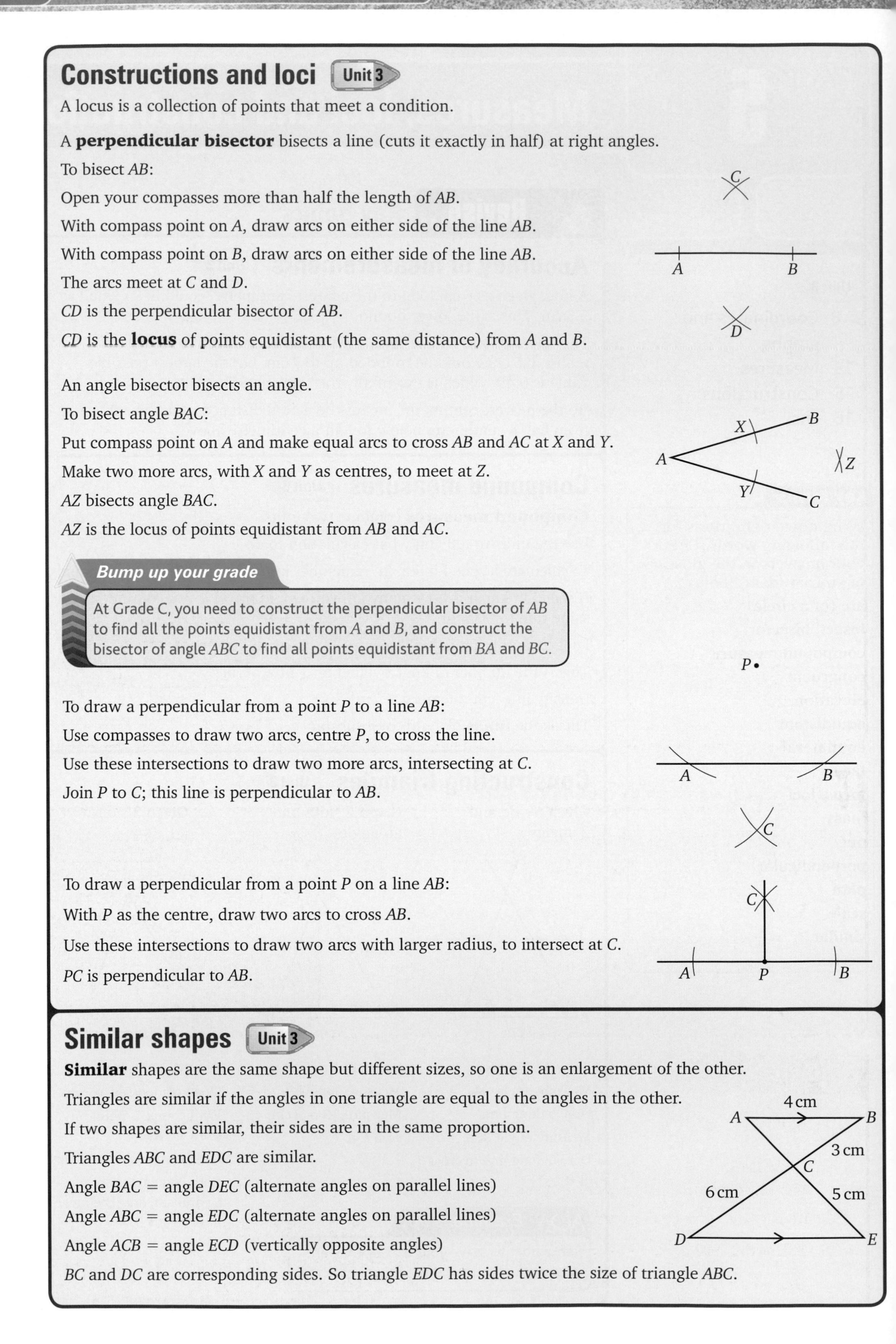

Constructions and loci

Unit 3

A locus is a collection of points that meet a condition.

A **perpendicular bisector** bisects a line (cuts it exactly in half) at right angles.

To bisect *AB*:

Open your compasses more than half the length of *AB*.

With compass point on *A*, draw arcs on either side of the line *AB*.

With compass point on *B*, draw arcs on either side of the line *AB*.

The arcs meet at *C* and *D*.

CD is the perpendicular bisector of *AB*.

CD is the **locus** of points equidistant (the same distance) from *A* and *B*.

An angle bisector bisects an angle.

To bisect angle *BAC*:

Put compass point on *A* and make equal arcs to cross *AB* and *AC* at *X* and *Y*.

Make two more arcs, with *X* and *Y* as centres, to meet at *Z*.

AZ bisects angle *BAC*.

AZ is the locus of points equidistant from *AB* and *AC*.

Bump up your grade

At Grade C, you need to construct the perpendicular bisector of *AB* to find all the points equidistant from *A* and *B*, and construct the bisector of angle *ABC* to find all points equidistant from *BA* and *BC*.

To draw a perpendicular from a point *P* to a line *AB*:

Use compasses to draw two arcs, centre *P*, to cross the line.

Use these intersections to draw two more arcs, intersecting at *C*.

Join *P* to *C*; this line is perpendicular to *AB*.

To draw a perpendicular from a point *P* on a line *AB*:

With *P* as the centre, draw two arcs to cross *AB*.

Use these intersections to draw two arcs with larger radius, to intersect at *C*.

PC is perpendicular to *AB*.

Similar shapes

Unit 3

Similar shapes are the same shape but different sizes, so one is an enlargement of the other.

Triangles are similar if the angles in one triangle are equal to the angles in the other.

If two shapes are similar, their sides are in the same proportion.

Triangles *ABC* and *EDC* are similar.

Angle *BAC* = angle *DEC* (alternate angles on parallel lines)

Angle *ABC* = angle *EDC* (alternate angles on parallel lines)

Angle *ACB* = angle *ECD* (vertically opposite angles)

BC and *DC* are corresponding sides. So triangle *EDC* has sides twice the size of triangle *ABC*.

Congruent shapes Unit 3

Congruent shapes are exactly the same shape and size.

There are four ways of proving triangles are congruent.

If you know that:

- both triangles have 3 corresponding sides equal [SSS], or
- both triangles have 2 angles and one corresponding side equal [ASA], or
- both triangles have 2 corresponding sides and the angle between them equal [SAS], or
- both triangles have a right angle, hypotenuse and another side equal [RHS],

the triangles must be congruent.

Aim higher

For a Grade A, learn the four proofs of congruence and apply them.

Plans and elevations Unit 3

This shape is made of 8 cubes.

The front **elevation**, side elevation and plan view are the views from directly in front, from the side and from above.

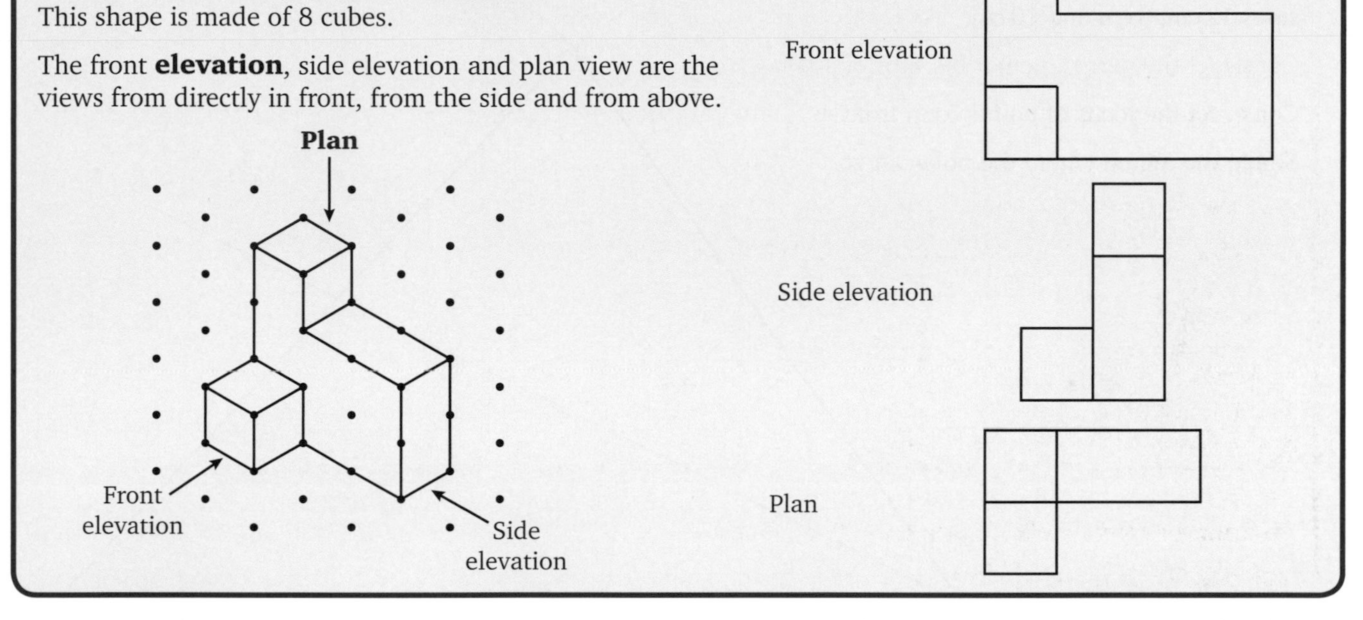

Example Accuracy of measurements Unit 3

A parcel weighs 12 g to the nearest gram.

What are the upper and lower bounds?

Solution

The parcel could weigh 0.5 g either side of the 12 g.

The lower bound is 11.5 g and the upper bound is 12.5 g.

AQA Examiner's tip

Some candidates are aware that 12.5 g would actually round up to 13 g, so they give the upper bound as 12.4 g. However, the mass of the parcel could be greater than that, for example it could be 12.49 g, so the upper bound has to be given as 12.5 g.

C

Example Compound measures Unit 3

A silver cube has sides of 3 cm. The density of silver is 10.5 g/cm^3. Calculate the mass of the cube.

A

Solution

The volume of the cube is 3 cm $\times$ 3 cm $\times$ 3 cm = 27 cm^3

Density is measured in g/cm^3.

So, for the cube, $\frac{\text{mass}}{27} = 10.5$

The mass = 10.5 $\times$ 27 = 283.5 g

AQA Examiner's tip

Make sure that you use the correct units. In this example you needed to find the volume in cm^3 first.

Example — Constructions and loci (Unit 3)

C

A triangular field *ABC* has sides of 120 m, 90 m and 100 m as shown.

A telegraph pole is to be placed closer to *AC* than *AB*, and more than 50 m from *A*.

Use a scale of 1 cm to 10 m to make a scale drawing and construct the possible position of the pole.

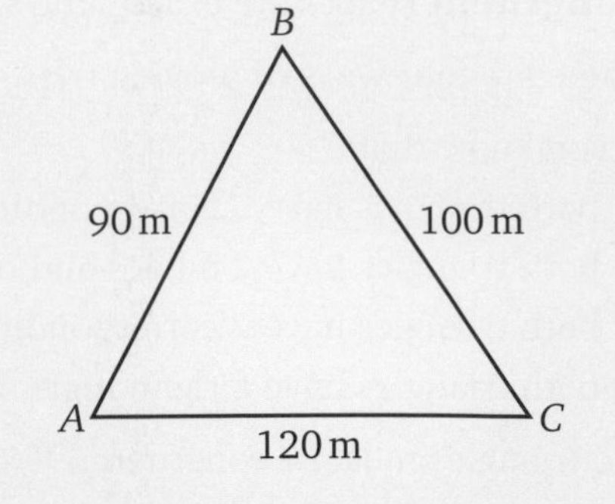

Solution

Use compasses to construct a triangle of sides 12 cm, 9 cm and 10 cm.

Construct the perpendicular bisector of angle *BAC*.

Construct the locus of points 5 cm from *A*.

Shade the region where the pole can go.

B

A

C

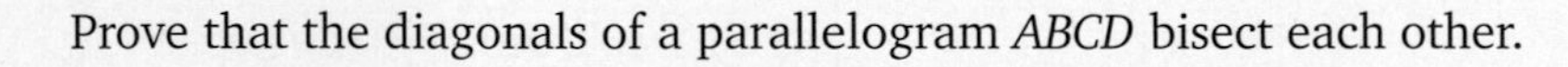

Example — Congruent triangles (Unit 3)

A

Prove that the diagonals of a parallelogram *ABCD* bisect each other.

A B E D C

Solution

In triangles *ABE*, *CDE*:

$AB = CD$ (opposite side of a parallelogram are equal)

$EAB = ECD$ (alternate angles on parallel lines)

$AEB = CED$ (vertically opposite angles)

Triangle *ABE* is congruent to triangle *CDE* (ASA)

So $AE = EC$, $BE = ED$

The diagonals bisect each other.

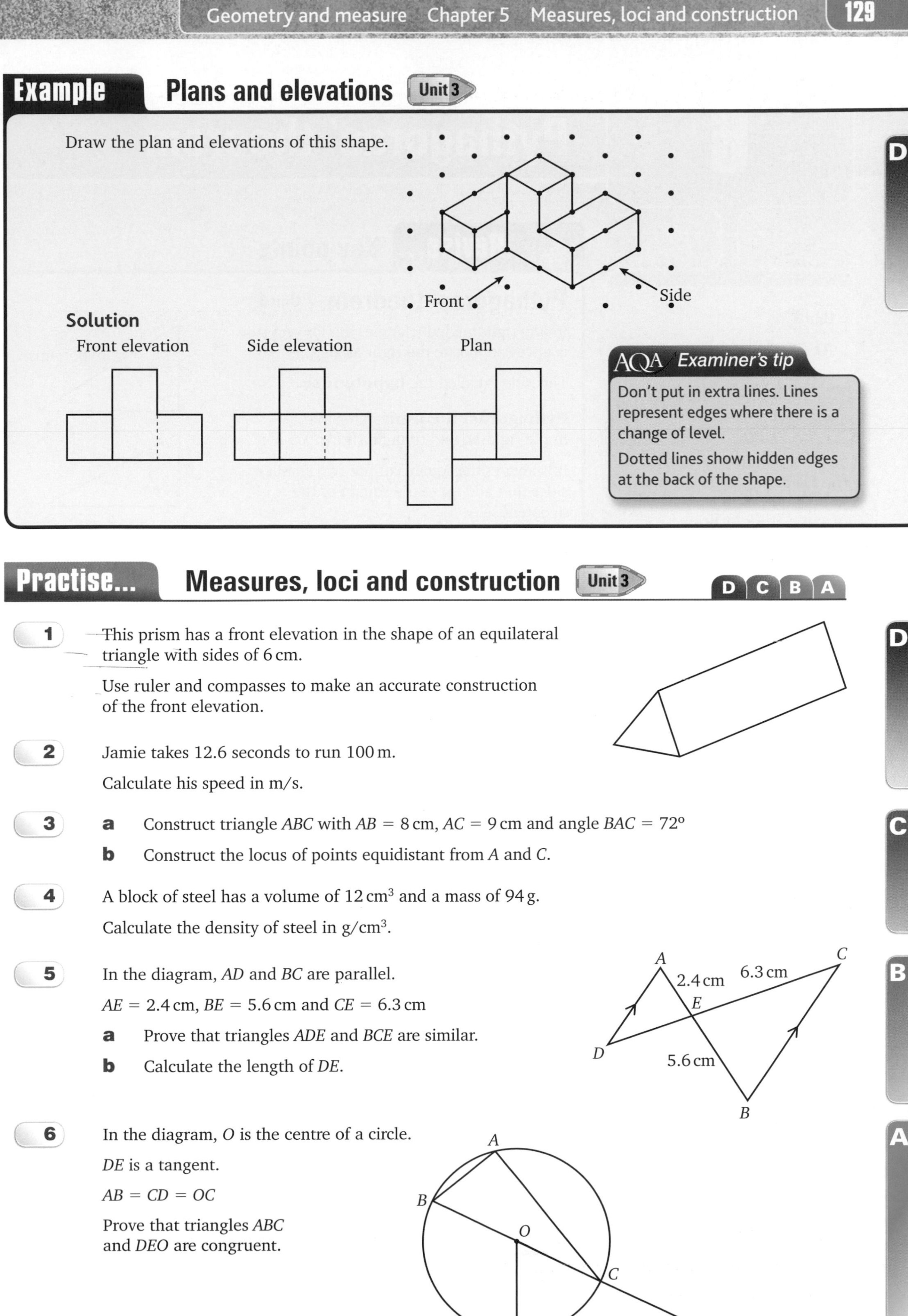

Example Plans and elevations Unit 3

Draw the plan and elevations of this shape.

Solution

AQA Examiner's tip

Don't put in extra lines. Lines represent edges where there is a change of level.

Dotted lines show hidden edges at the back of the shape.

Practise... Measures, loci and construction Unit 3

D C B A

1 This prism has a front elevation in the shape of an equilateral triangle with sides of 6 cm.

Use ruler and compasses to make an accurate construction of the front elevation.

2 Jamie takes 12.6 seconds to run 100 m.

Calculate his speed in m/s.

3

a Construct triangle ABC with $AB = 8$ cm, $AC = 9$ cm and angle $BAC = 72°$

b Construct the locus of points equidistant from A and C.

4 A block of steel has a volume of 12 cm^3 and a mass of 94 g.

Calculate the density of steel in g/cm^3.

5 In the diagram, AD and BC are parallel.

$AE = 2.4$ cm, $BE = 5.6$ cm and $CE = 6.3$ cm

a Prove that triangles ADE and BCE are similar.

b Calculate the length of DE.

6 In the diagram, O is the centre of a circle.

DE is a tangent.

$AB = CD = OC$

Prove that triangles ABC and DEO are congruent.

Geometry and measure

6 Pythagoras' theorem

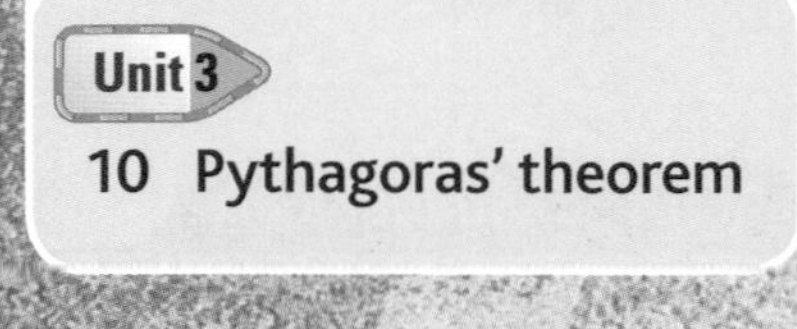

Key terms

Write down definitions for the following words. Check your answers in the glossary of your Student Book.

hypotenuse

Pythagoras' theorem

Revise... Key points

Pythagoras' theorem (Unit 3)

In any right-angled triangle, the longest side is always opposite the right angle.

This side is called the **hypotenuse**.

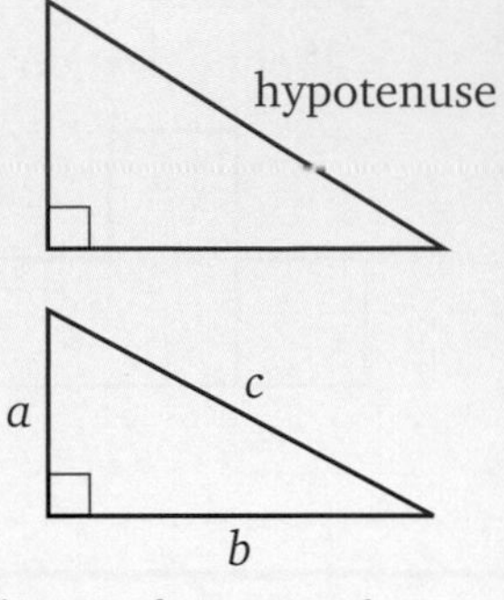

Pythagoras' theorem states that, in the right-angled triangle shown, $a^2 + b^2 = c^2$

It is always the squares of the two smaller sides that add up to the square of the hypotenuse.

Right-angled triangles can be formed from other shapes, for example isosceles and equilateral triangles, and diagonals of rectangles. Wherever a right-angled triangle is formed you can use Pythagoras' theorem to find the length of the third side.

AQA *Examiner's tip*

When finding the hypotenuse, you must add the squares of the other two sides but when finding a shorter side you must find the difference between the squares of the other two sides.

Bump up your grade

For a Grade C, you need to use Pythagoras' theorem in right-angled triangles, to find the hypotenuse and also to find one of the shorter sides.

Pythagoras in three dimensions (Unit 3)

Pythagoras' theorem can be used in three dimensions.

This often involves two calculations.

To find the length of the diagonal BH:

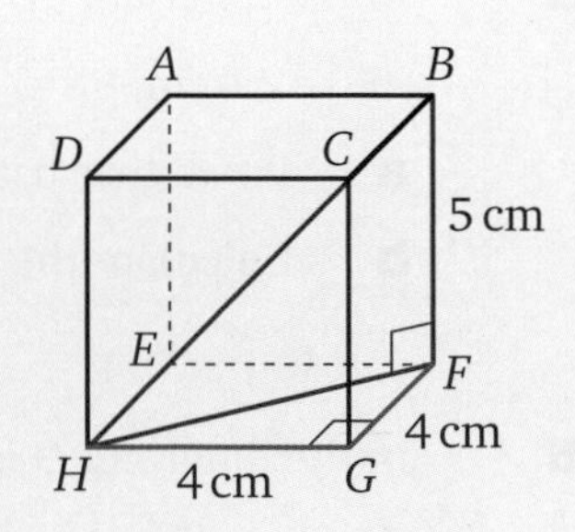

Use the blue right-angled triangle to calculate the length of FH:

$GH^2 + GF^2 = FH^2$

$16 + 16 = FH^2$

$FH = \sqrt{32}$

Use the red right-angled triangle to calculate BH:

$FH^2 + BF^2 = BH^2$

$(\sqrt{32})^2 + 5^2 = BH^2$

$32 + 25 = BH^2$

$BH = \sqrt{57} = 7.5$ cm (to 1 d.p.)

AQA *Examiner's tip*

There is no need to work out $\sqrt{32}$, as you need to square it again in the second calculation.

Example Pythagoras' theorem Unit 3

C

A is the point (4, 1) and B is the point (2, −3).

Calculate the length of AB.

AQA Examiner's tip

When finding the hypotenuse, you must add the squares of the other two sides, but when finding a shorter side, you must find the difference between the squares of the other two sides.

Solution

Draw a sketch:

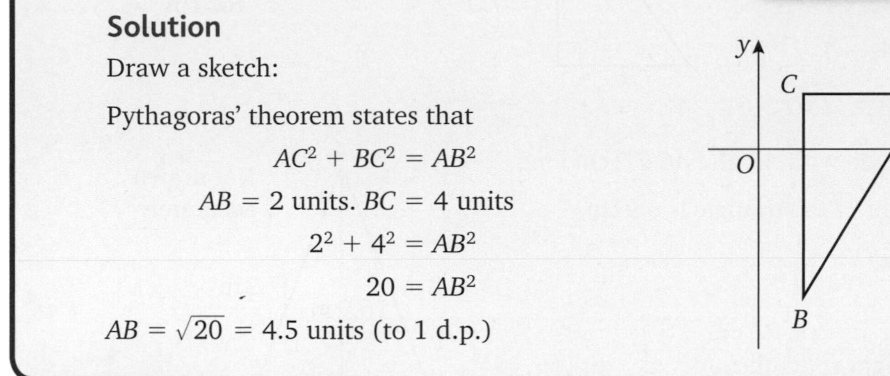

Pythagoras' theorem states that

$$AC^2 + BC^2 = AB^2$$

$AB = 2$ units. $BC = 4$ units

$$2^2 + 4^2 = AB^2$$

$$20 = AB^2$$

$AB = \sqrt{20} = 4.5$ units (to 1 d.p.)

Example Pythagoras in three dimensions Unit 3

A

A square-based pyramid $ABCDE$ has a square base with sides of 7 cm.

Calculate the height, EX, of the pyramid.

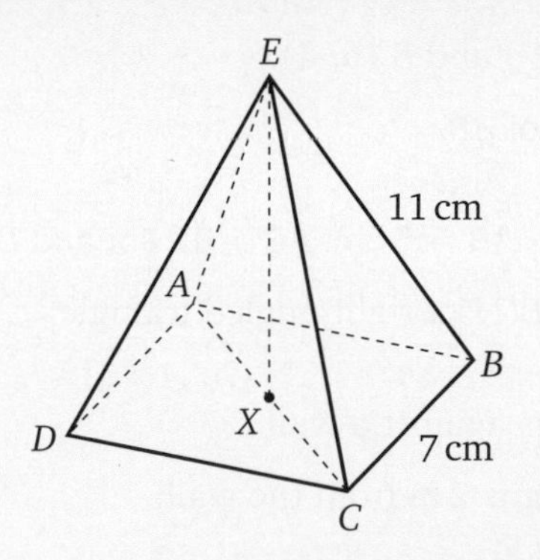

Solution

Draw the square base $ABCD$ and the diagonals.

$$AB^2 + BC^2 = AC^2$$

$$7^2 + 7^2 = AC^2$$

$$98 = AC^2$$

$$AC = \sqrt{98}$$

$$CX = \frac{\sqrt{98}}{2}$$

Hint

There is no need to work this out, as you will square it at the next stage.

Now draw triangle EXC:

$$EX^2 + XC^2 = EC^2$$

$$EX^2 + \left(\frac{\sqrt{98}}{2}\right)^2 = 11^2$$

$$EX^2 + \frac{98}{4} = 121$$

$$EX^2 = 96.5$$

$EX = 9.8$ cm (to 1 d.p.)

Practise... Pythagoras' theorem Unit 3

C B A

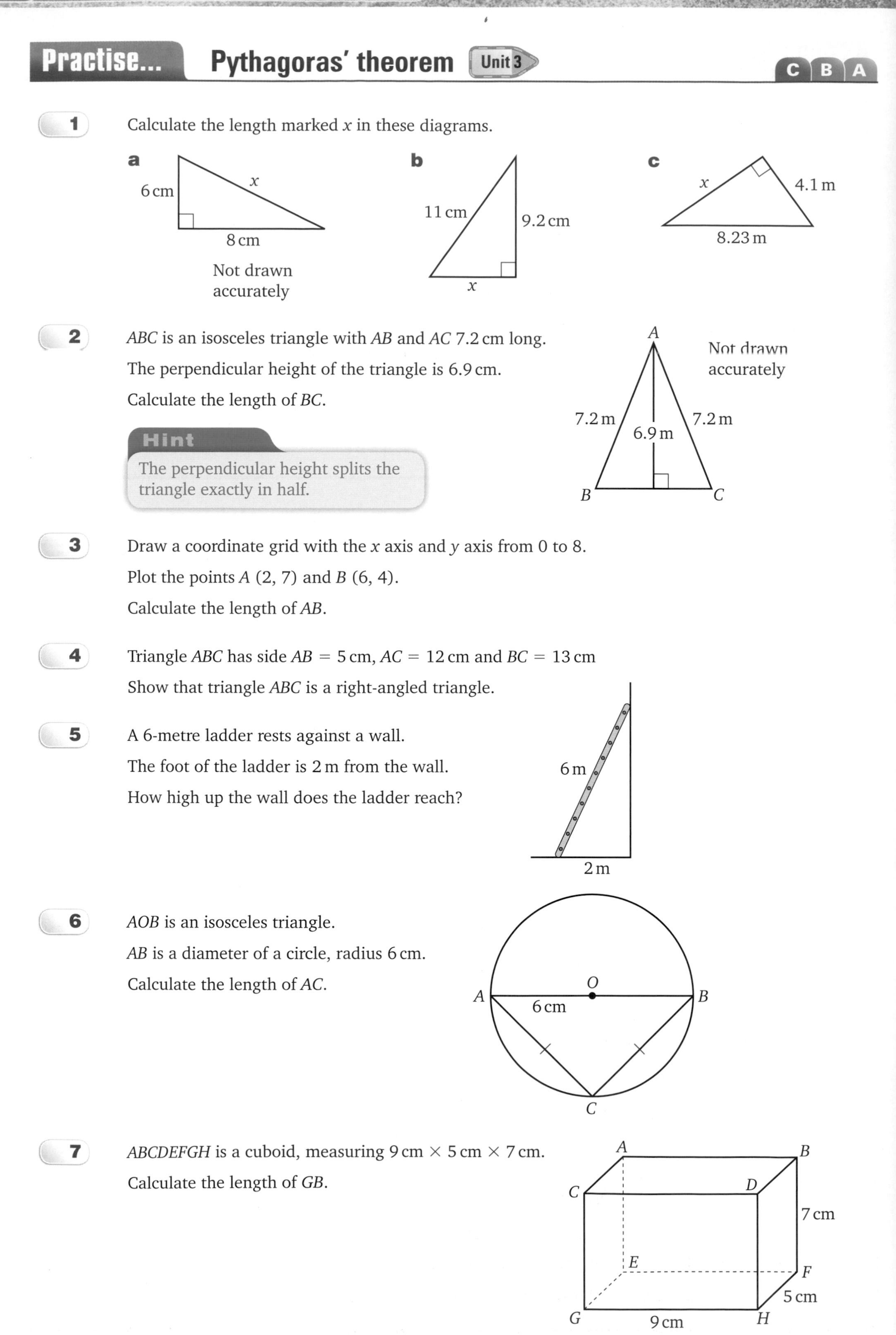

C

1 Calculate the length marked x in these diagrams.

a 6 cm, x, 8 cm

Not drawn accurately

b 11 cm, 9.2 cm, x

c x, 4.1 m, 8.23 m

2 *ABC* is an isosceles triangle with *AB* and *AC* 7.2 cm long.

The perpendicular height of the triangle is 6.9 cm.

Calculate the length of *BC*.

Hint

The perpendicular height splits the triangle exactly in half.

3 Draw a coordinate grid with the x axis and y axis from 0 to 8.

Plot the points *A* (2, 7) and *B* (6, 4).

Calculate the length of *AB*.

4 Triangle *ABC* has side $AB = 5$ cm, $AC = 12$ cm and $BC = 13$ cm

Show that triangle *ABC* is a right-angled triangle.

5 A 6-metre ladder rests against a wall.

The foot of the ladder is 2 m from the wall.

How high up the wall does the ladder reach?

B

6 *AOB* is an isosceles triangle.

AB is a diameter of a circle, radius 6 cm.

Calculate the length of *AC*.

A

7 *ABCDEFGH* is a cuboid, measuring 9 cm × 5 cm × 7 cm.

Calculate the length of *GB*.

Geometry and measure

7 Trigonometry

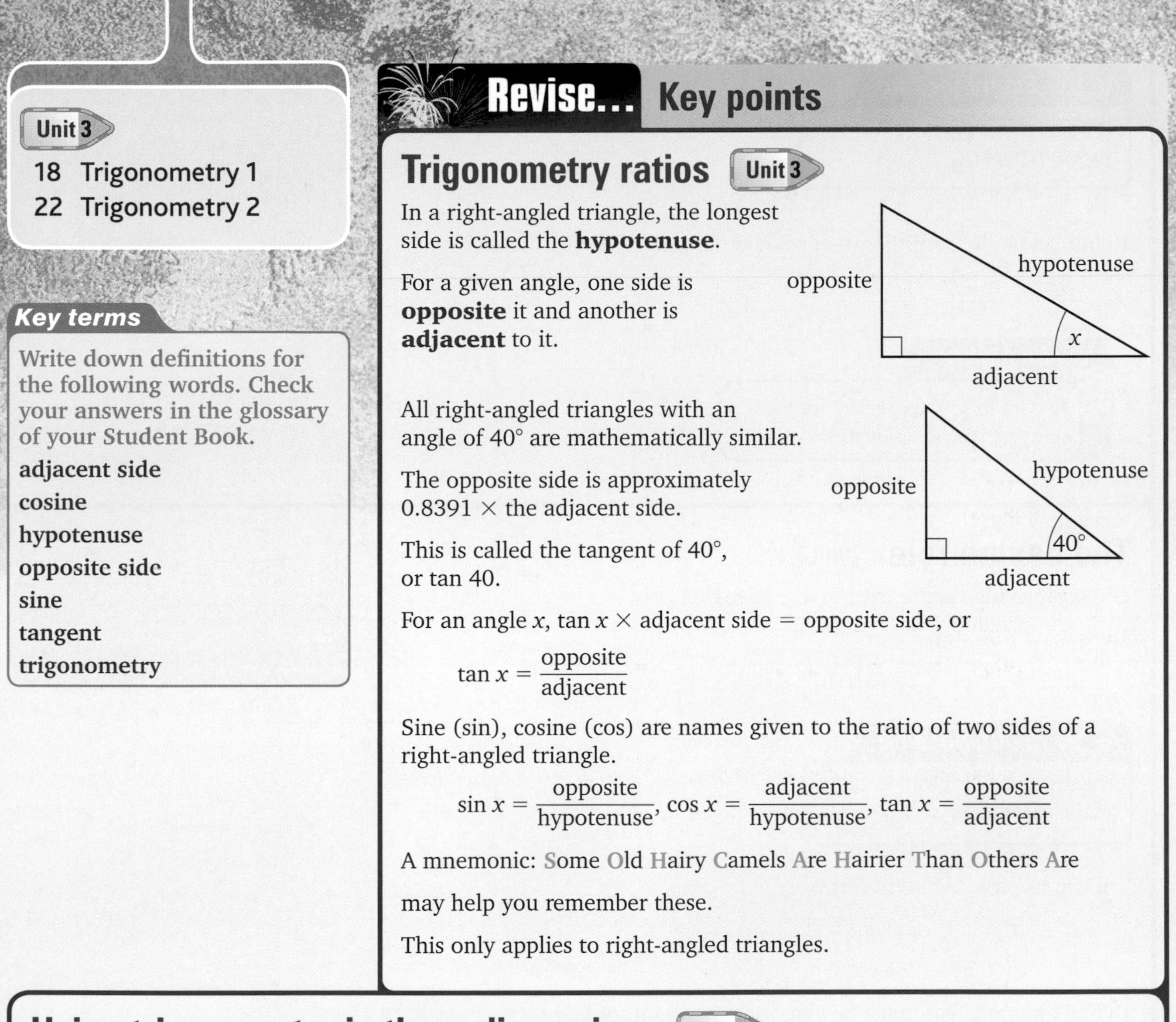

Unit 3

Key terms

Write down definitions for the following words. Check your answers in the glossary of your Student Book.

adjacent side
cosine
hypotenuse
opposite side
sine
tangent
trigonometry

Revise... Key points

Trigonometry ratios (Unit 3)

In a right-angled triangle, the longest side is called the **hypotenuse**.

For a given angle, one side is **opposite** it and another is **adjacent** to it.

All right-angled triangles with an angle of 40° are mathematically similar.

The opposite side is approximately 0.8391 × the adjacent side.

This is called the tangent of 40°, or tan 40.

For an angle x, tan x × adjacent side = opposite side, or

$$\tan x = \frac{\text{opposite}}{\text{adjacent}}$$

Sine (sin), cosine (cos) are names given to the ratio of two sides of a right-angled triangle.

$$\sin x = \frac{\text{opposite}}{\text{hypotenuse}}, \cos x = \frac{\text{adjacent}}{\text{hypotenuse}}, \tan x = \frac{\text{opposite}}{\text{adjacent}}$$

A mnemonic: Some Old Hairy Camels Are Hairier Than Others Are

may help you remember these.

This only applies to right-angled triangles.

Using trigonometry in three dimensions (Unit 3)

Trigonometry can be used in three dimensions. This often requires two calculations, sometimes including Pythagoras.

Imagine a straw inside a container with a square base.

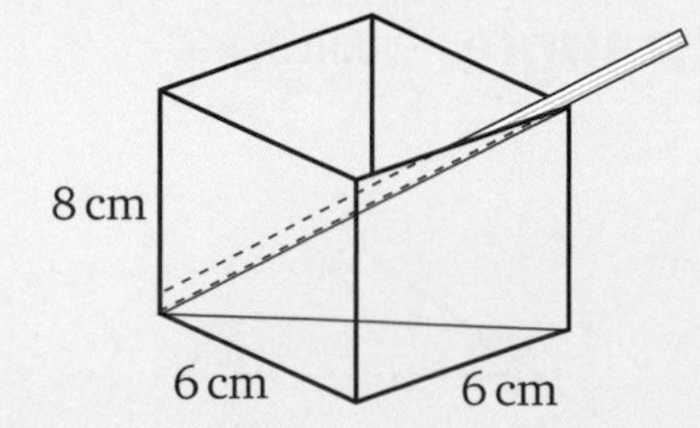

To calculate the angle between the straw and the base:

First, use Pythagoras' theorem to calculate the diagonal of the base.

Then use the right-angled triangle shown in red to calculate the angle.

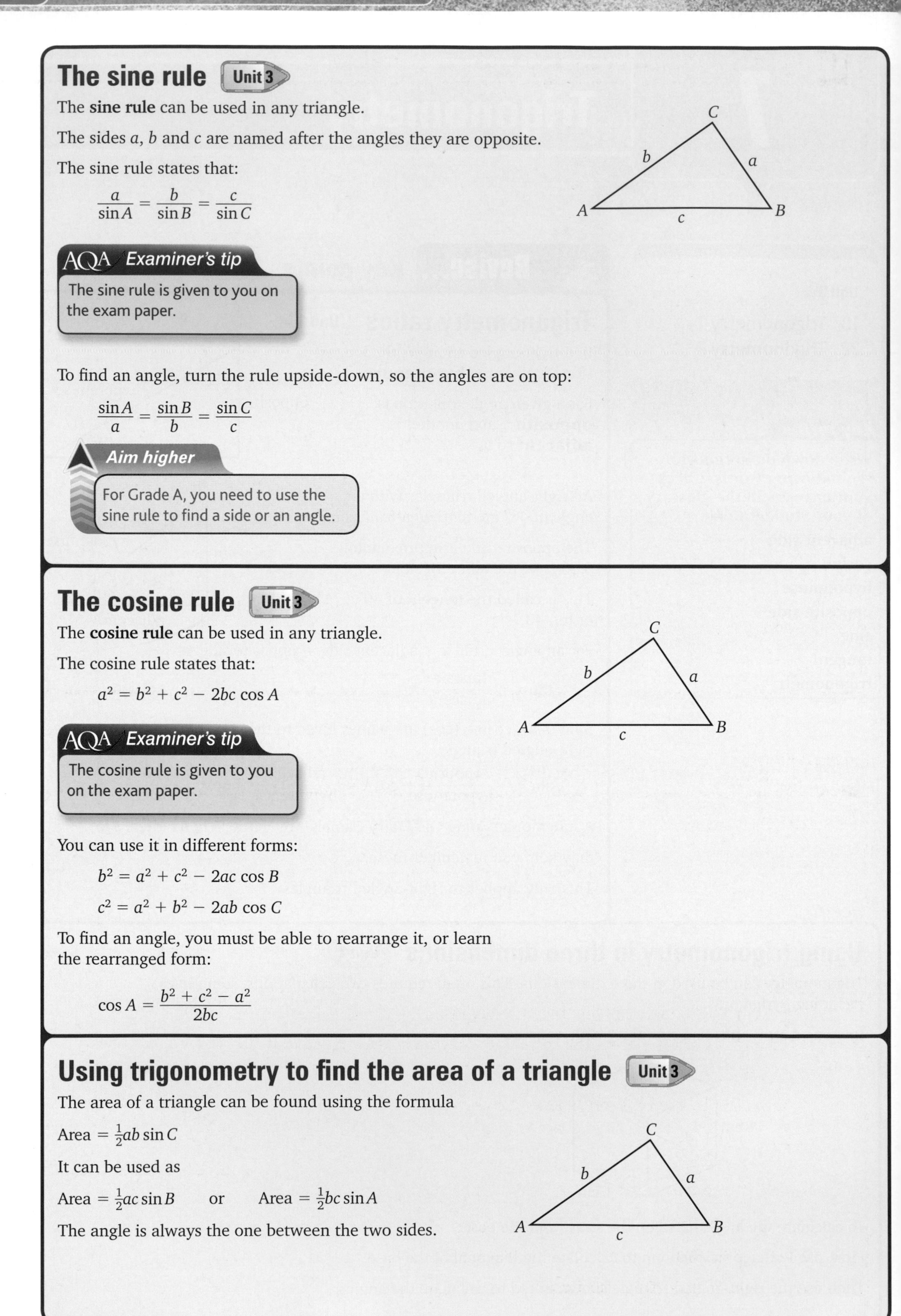

The sine rule Unit 3

The **sine rule** can be used in any triangle.

The sides a, b and c are named after the angles they are opposite.

The sine rule states that:

$$\frac{a}{\sin A} = \frac{b}{\sin B} = \frac{c}{\sin C}$$

AQA Examiner's tip

The sine rule is given to you on the exam paper.

To find an angle, turn the rule upside-down, so the angles are on top:

$$\frac{\sin A}{a} = \frac{\sin B}{b} = \frac{\sin C}{c}$$

Aim higher

For Grade A, you need to use the sine rule to find a side or an angle.

The cosine rule Unit 3

The **cosine rule** can be used in any triangle.

The cosine rule states that:

$$a^2 = b^2 + c^2 - 2bc \cos A$$

AQA Examiner's tip

The cosine rule is given to you on the exam paper.

You can use it in different forms:

$$b^2 = a^2 + c^2 - 2ac \cos B$$

$$c^2 = a^2 + b^2 - 2ab \cos C$$

To find an angle, you must be able to rearrange it, or learn the rearranged form:

$$\cos A = \frac{b^2 + c^2 - a^2}{2bc}$$

Using trigonometry to find the area of a triangle Unit 3

The area of a triangle can be found using the formula

Area $= \frac{1}{2}ab \sin C$

It can be used as

Area $= \frac{1}{2}ac \sin B$ or Area $= \frac{1}{2}bc \sin A$

The angle is always the one between the two sides.

Example Finding a side Unit 3

B

Using trigonometry to find a side.

Calculate the length of BC.

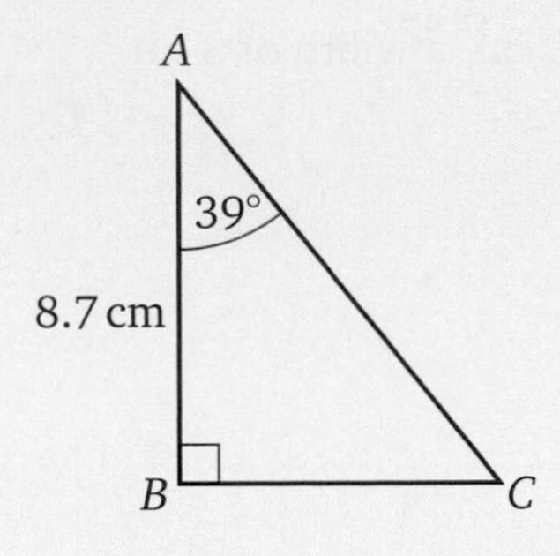

Solution

This is a right-angled triangle.
Label the diagram with hypotenuse, opposite and adjacent.

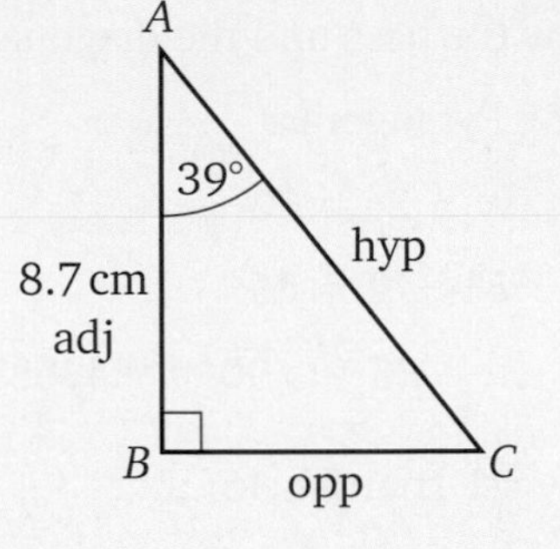

AQA Examiner's tip

Mark the 'hyp' first. Once you know which acute angle you are using (39°), mark the opposite side to that angle 'opp', which leaves the remaining side as 'adj'.

You know the adjacent, you want the opposite, so choose tangent.

$\tan x = \dfrac{\text{opposite}}{\text{adjacent}}$

$\tan 39 = \dfrac{BC}{8.7}$

$0.809784033 \times 8.7 = BC$

$BC = 7.0\,\text{cm}$ (to 1 d.p.)

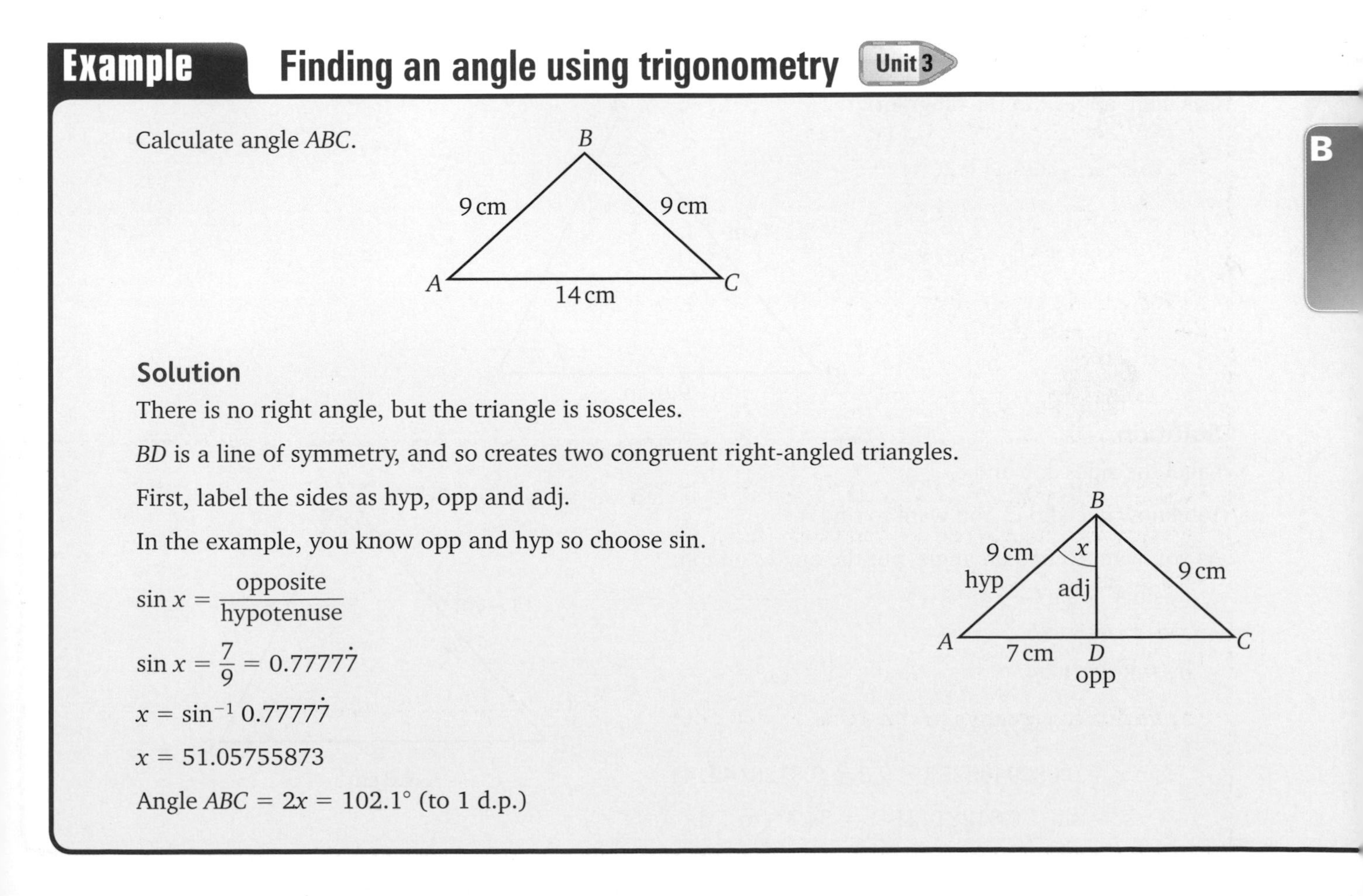

Example Finding an angle using trigonometry Unit 3

B

Calculate angle ABC.

Solution

There is no right angle, but the triangle is isosceles.

BD is a line of symmetry, and so creates two congruent right-angled triangles.

First, label the sides as hyp, opp and adj.

In the example, you know opp and hyp so choose sin.

$\sin x = \dfrac{\text{opposite}}{\text{hypotenuse}}$

$\sin x = \dfrac{7}{9} = 0.7777\dot{7}$

$x = \sin^{-1} 0.7777\dot{7}$

$x = 51.05755873$

Angle $ABC = 2x = 102.1°$ (to 1 d.p.)

Example Trigonometry in three dimensions Unit 3

B

A cuboid *ABCDEFGH* has a length of 9 cm, a width of 3 cm and a height of 5 cm.

Calculate angle *BGF*.

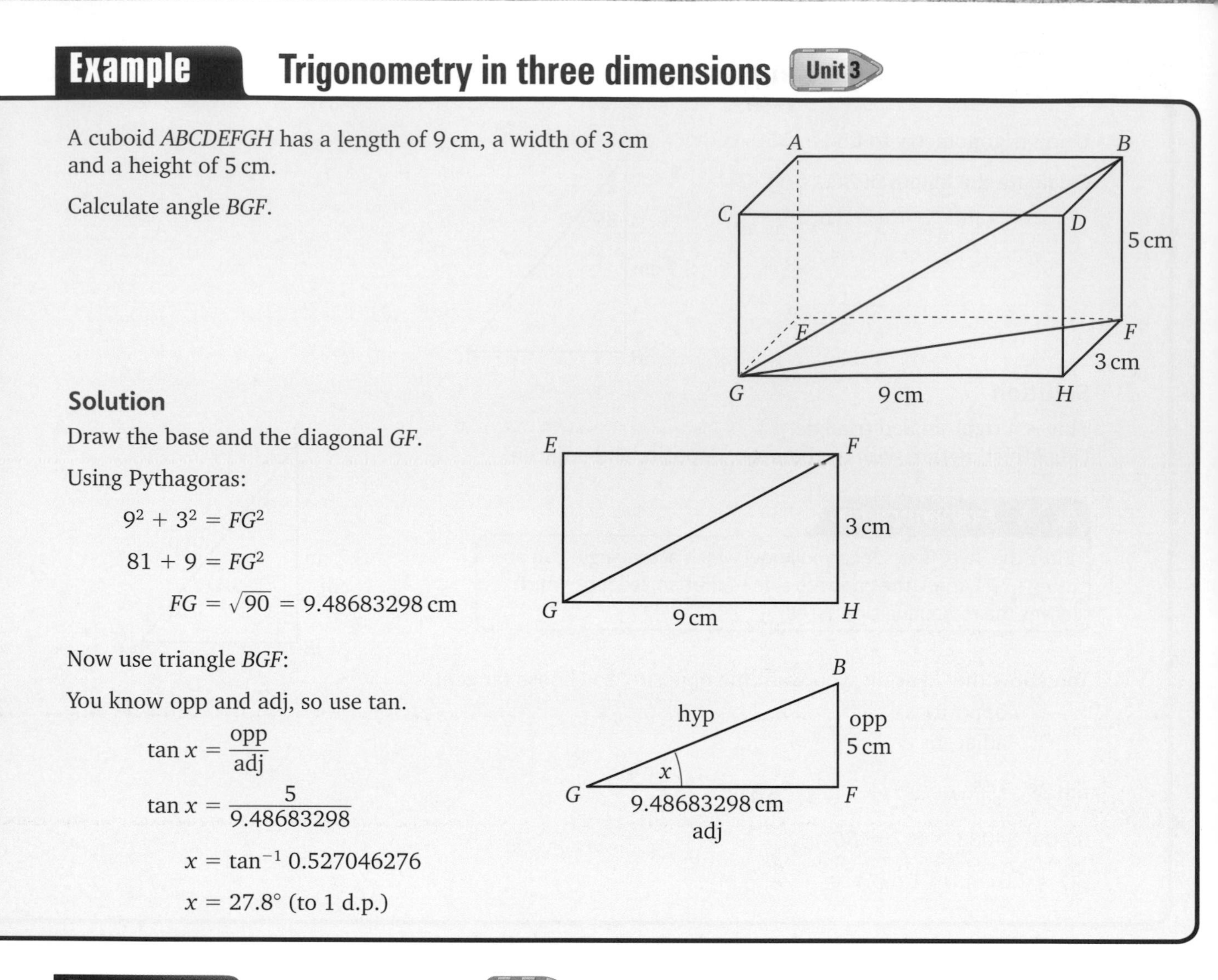

Solution

Draw the base and the diagonal *GF*.

Using Pythagoras:

$$9^2 + 3^2 = FG^2$$

$$81 + 9 = FG^2$$

$$FG = \sqrt{90} = 9.48683298 \text{ cm}$$

Now use triangle *BGF*:

You know opp and adj, so use tan.

$$\tan x = \frac{\text{opp}}{\text{adj}}$$

$$\tan x = \frac{5}{9.48683298}$$

$$x = \tan^{-1} 0.527046276$$

$$x = 27.8° \text{ (to 1 d.p.)}$$

Example The sine rule Unit 3

A

Calculate angle x in the diagram.

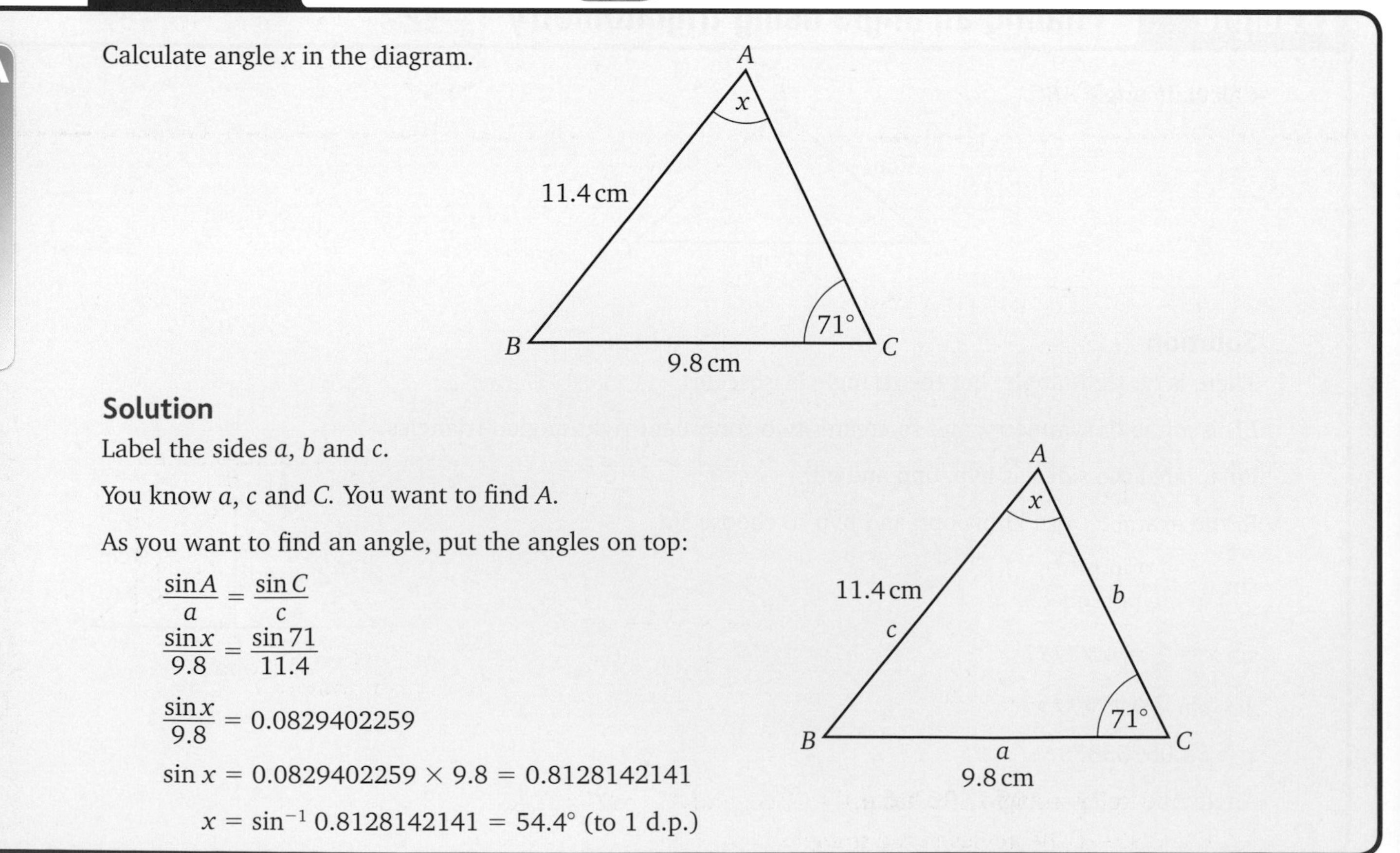

Solution

Label the sides a, b and c.

You know a, c and C. You want to find A.

As you want to find an angle, put the angles on top:

$$\frac{\sin A}{a} = \frac{\sin C}{c}$$

$$\frac{\sin x}{9.8} = \frac{\sin 71}{11.4}$$

$$\frac{\sin x}{9.8} = 0.0829402259$$

$$\sin x = 0.0829402259 \times 9.8 = 0.8128142141$$

$$x = \sin^{-1} 0.8128142141 = 54.4° \text{ (to 1 d.p.)}$$

Example The cosine rule Unit 3

A

A, B and C are three towns.

B is on a bearing of 072° from A.

The distances between the towns are shown on the diagram.

Calculate the bearing of C from A.

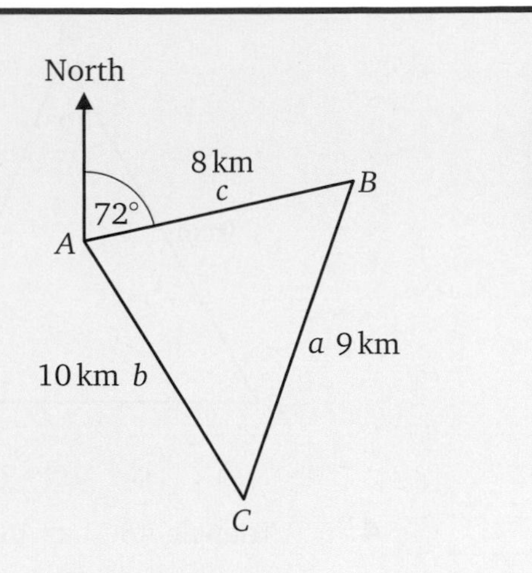

Solution

To calculate angle A:

$a^2 = b^2 + c^2 - 2bc \cos A$

$9^2 = 10^2 + 8^2 - 2 \times 10 \times 8 \cos A$

$81 = 164 - 160 \cos A$

$160 \cos A = 164 - 81$

$\cos A = \dfrac{83}{160} = 0.51875$

$A = \cos^{-1} 0.51875 = 58.75155874°$

Bearing is $72° + 58.75155874° = 131°$ to nearest degree.

Example Area of a triangle Unit 3

A

Calculate the area of triangle ABC.

Solution

You know angle B, so use the formula

Area $= \frac{1}{2}ac \sin B$

Area $= \frac{1}{2} \times 7.2 \times 6.8 \times \sin 52$

Area $= 19.3\ \text{cm}^2$ (to 1 d.p.)

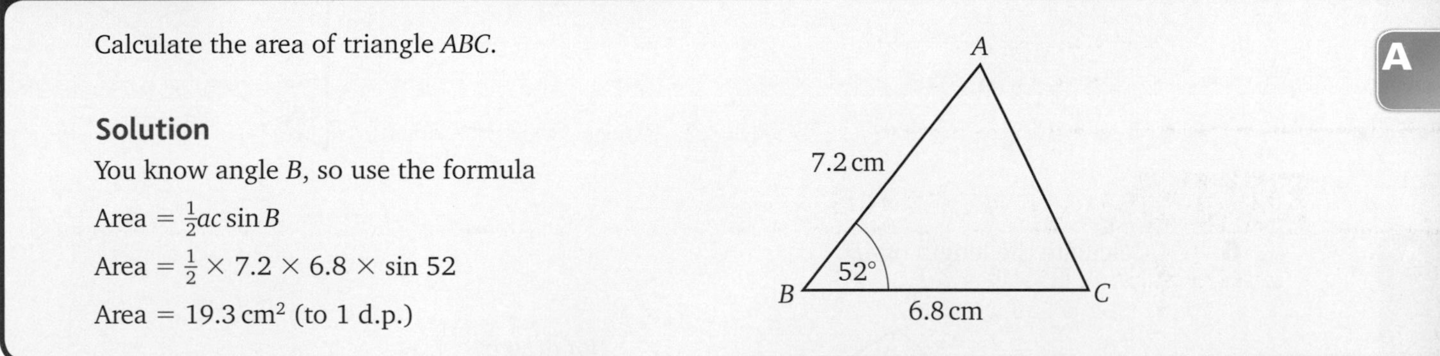

Practise... Trigonometry Unit 3

B A A*

B

1 Calculate the sides and angles marked with letters.

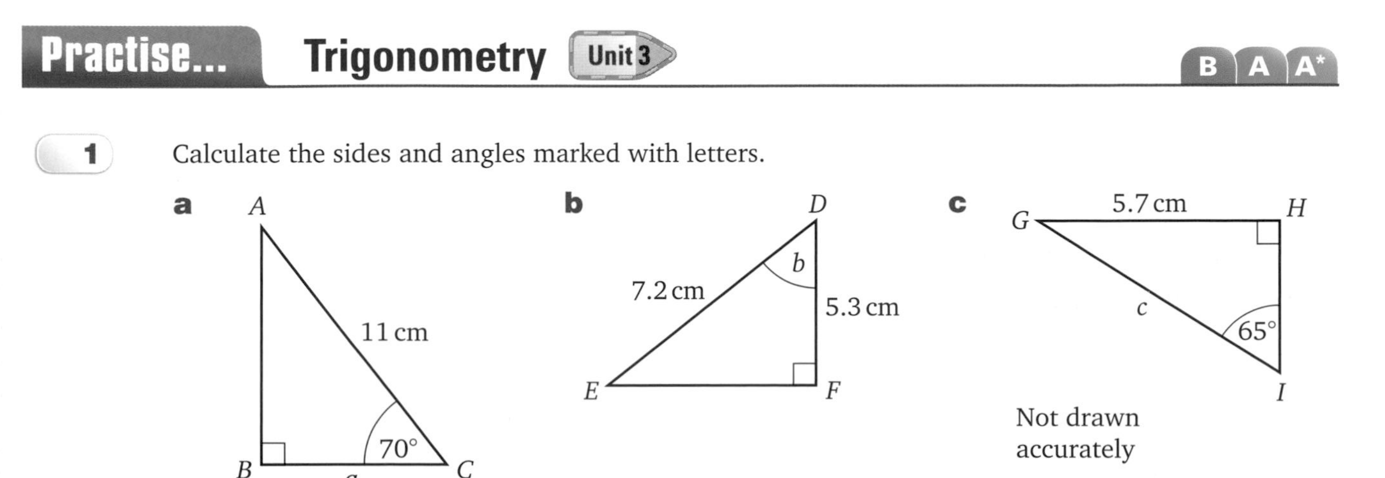

2 An isosceles triangle ABC has $AB = 12$ cm and $AC = BC = 7$ cm. Calculate all the angles in the triangle.

A

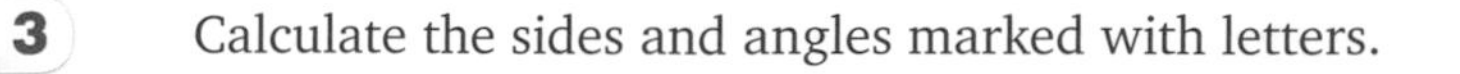

3 Calculate the sides and angles marked with letters.

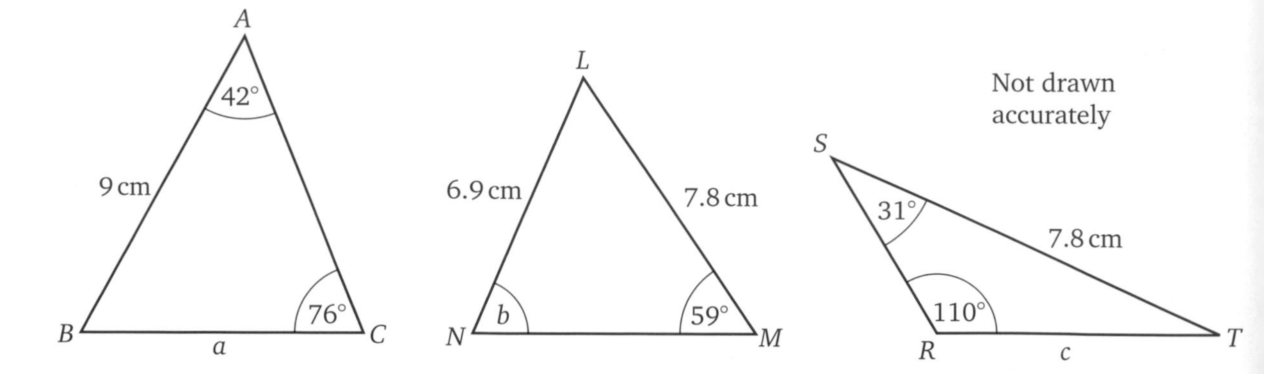

4 Triangle ABC has $AB = 9.2$ cm, $BC = 7.4$ cm and angle $ABC = 56°$.

Calculate the area of triangle ABC.

5 Y is 4.4 km due south of X.

Z is on a bearing of 068° from Y and 113° from X.

Calculate the distance from X to Z, and from Y to Z.

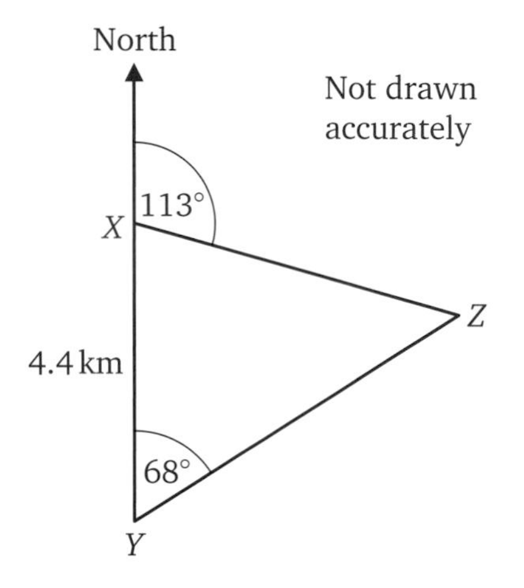

A*

6 Calculate the length of AD.

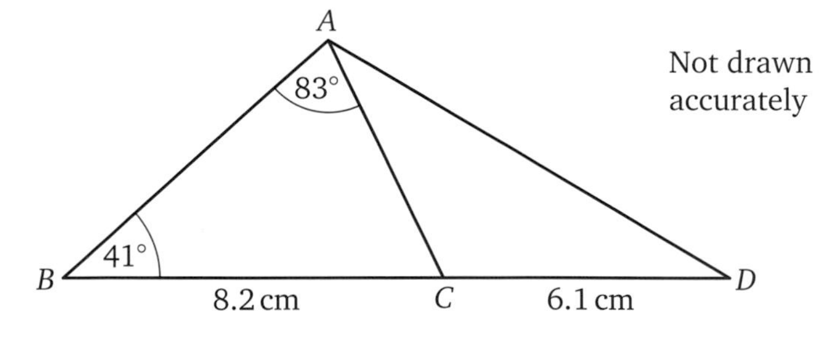

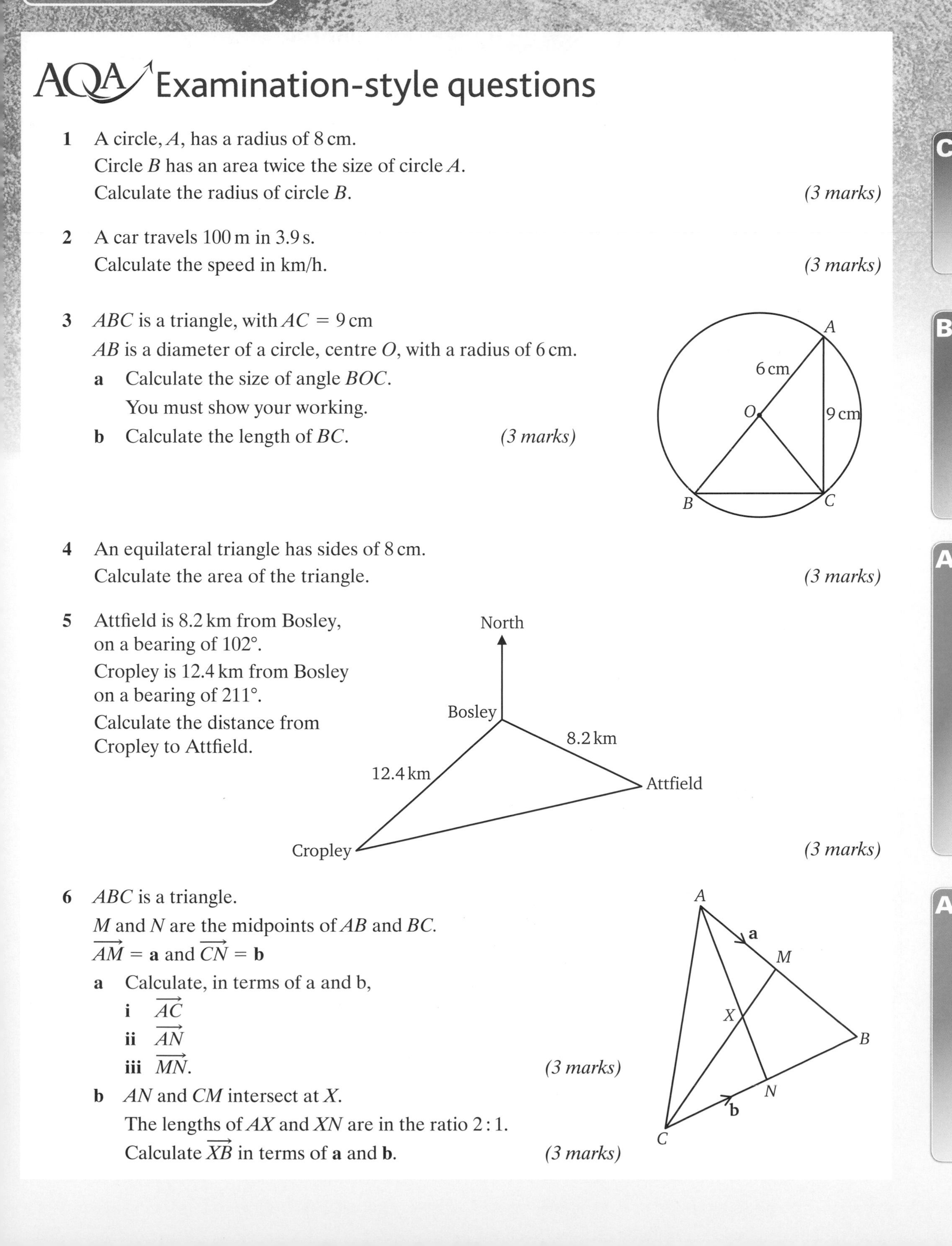

AQA Examination-style questions

1 A circle, A, has a radius of 8 cm.
Circle B has an area twice the size of circle A.
Calculate the radius of circle B. *(3 marks)*

2 A car travels 100 m in 3.9 s.
Calculate the speed in km/h. *(3 marks)*

3 ABC is a triangle, with $AC = 9$ cm
AB is a diameter of a circle, centre O, with a radius of 6 cm.

a Calculate the size of angle BOC.
You must show your working.

b Calculate the length of BC. *(3 marks)*

4 An equilateral triangle has sides of 8 cm.
Calculate the area of the triangle. *(3 marks)*

5 Attfield is 8.2 km from Bosley, on a bearing of 102°.
Cropley is 12.4 km from Bosley on a bearing of 211°.
Calculate the distance from Cropley to Attfield. *(3 marks)*

6 ABC is a triangle.
M and N are the midpoints of AB and BC.
$\overrightarrow{AM} = \mathbf{a}$ and $\overrightarrow{CN} = \mathbf{b}$

a Calculate, in terms of a and b,

i $\overrightarrow{AC}$

ii $\overrightarrow{AN}$

iii $\overrightarrow{MN}$. *(3 marks)*

b AN and CM intersect at X.
The lengths of AX and XN are in the ratio 2 : 1.
Calculate $\overrightarrow{XB}$ in terms of **a** and **b**. *(3 marks)*

Essential skills

1 Problem-solving

All Units 1 2 3

Statistics and number
Algebra and number
Algebra and geometry

Key terms

These terms will be used in your exam. Make sure you understand what they mean, and write your own definitions.

All Units 1 2 3

Explain
Explain why
Give a reason for your answer
Show
Show how you decide
Show that
Show working to justify your answer
Show working to support your answer
You must show your working

Compare data sets
Test a hypothesis

Units 2 3

Set up and solve equations

Use algebra to support and construct arguments

Revise... Key points

Questions in your maths exams are not always about a particular piece of mathematics that you have learnt.

In some questions you have to use the maths you know to solve a problem about mathematics or real life.

In these questions you are not told the mathematical method you have to use. You have to choose it yourself. You have to stop, think and try to puzzle out what you need to do.

Here are some of the things you might try.

- Working systematically

  This means being methodical and organised.

  It also means trying to break up a problem into its separate parts and then doing one part at a time.

- Working backwards

  This means trying out inverse operations using a value given in the problem to see whether it helps.

- Finding examples that fit

  This is useful whenever an answer has to fit more than one condition. Start by trying a value that fits one of the conditions and see whether it fits any of the others. Then change the value systematically until it fits all the conditions.

- Finding a relationship

  Try out connections between values given in the question and see whether they help.

- Choosing the maths

  This means choosing the maths you could use to help you solve the problem, then trying it to see whether it does.

Aim higher

You will need to be able to answer complex problem-solving questions to get a Grade A or A*.

AQA Examiner's tip

Sometimes problem-solving questions have more words than other questions. So it is important to read the question carefully, a sentence at a time, and try to pick out the important pieces of information.

Do this a few times until you are sure what the question is asking you to do.

When you have finished your answer, read the question again to make sure that you have done all that it has asked.

Example Unit 1

B

The box plots show the test scores of the boys and girls in an end-of-year maths test.

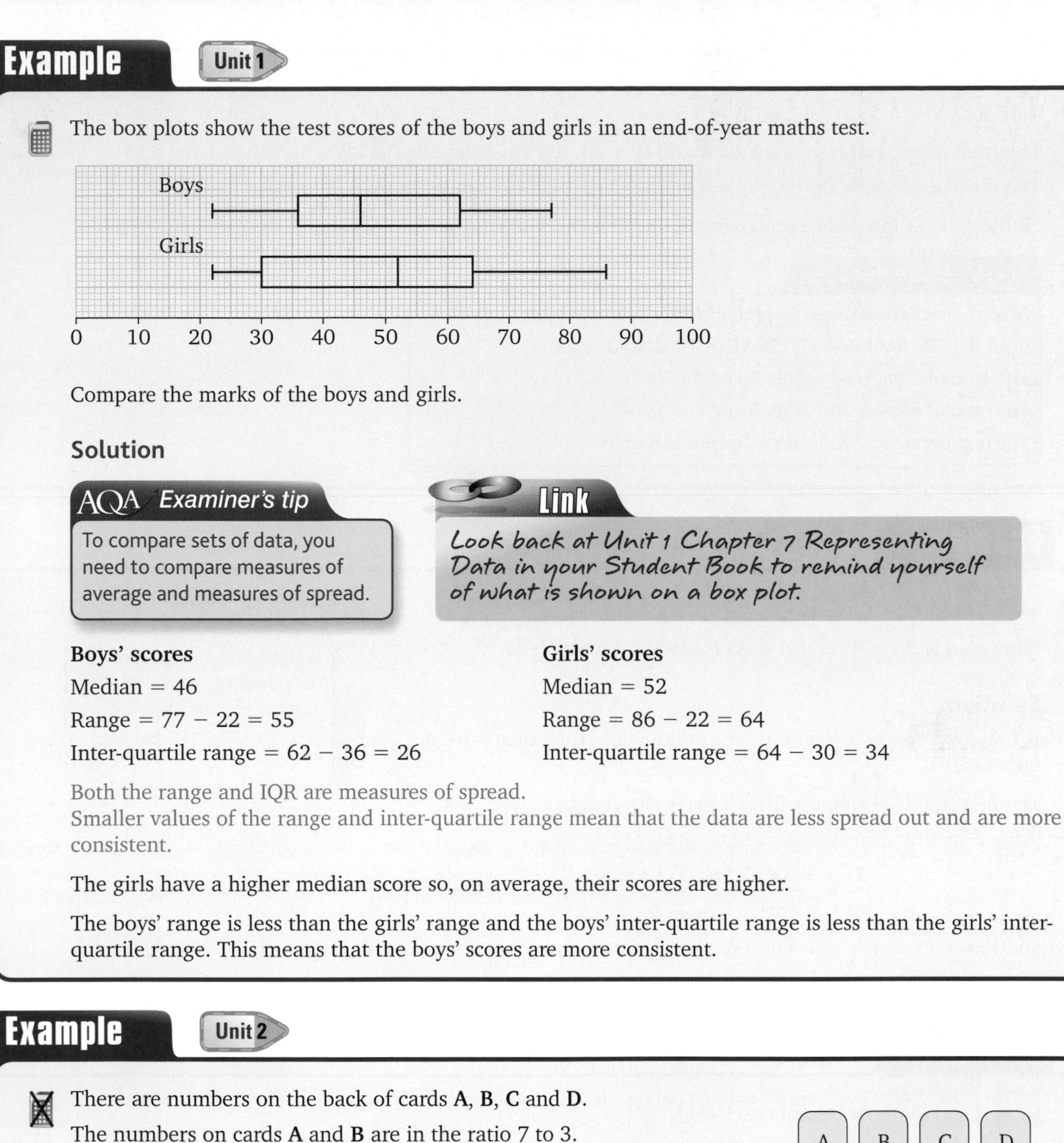

Compare the marks of the boys and girls.

Solution

AQA Examiner's tip

To compare sets of data, you need to compare measures of average and measures of spread.

Link

Look back at Unit 1 Chapter 7 Representing Data in your Student Book to remind yourself of what is shown on a box plot.

Boys' scores

Median = 46

Range = 77 − 22 = 55

Inter-quartile range = 62 − 36 = 26

Girls' scores

Median = 52

Range = 86 − 22 = 64

Inter-quartile range = 64 − 30 = 34

Both the range and IQR are measures of spread.
Smaller values of the range and inter-quartile range mean that the data are less spread out and are more consistent.

The girls have a higher median score so, on average, their scores are higher.

The boys' range is less than the girls' range and the boys' inter-quartile range is less than the girls' inter-quartile range. This means that the boys' scores are more consistent.

Example Unit 2

A

There are numbers on the back of cards **A**, **B**, **C** and **D**.

The numbers on cards **A** and **B** are in the ratio 7 to 3.

The number on card **C** is 10 less than the number on card **A**.

The number on card **D** is 30 more than the number on card **B**.

The numbers on cards **C** and **D** are in the ratio 3 : 7

Show that the numbers on cards **A** and **D** are in the same ratio as the numbers on cards **B** and **C**.

A B C D

Solution

Let the numbers on cards **A** and **B** be $7x$ and $3x$.

This is the most straightforward way of writing any two numbers in the ratio 7 : 3

So, the number on card **C** = $7x - 10$ and the number on card **D** = $3x + 30$

So $(7x - 10) : (3x + 30) = 3 : 7$

So $\frac{7x - 10}{3x + 25} = \frac{3}{7}$

So $7(7x - 10) = 3(3x + 30)$

$49x - 70 = 9x + 90$

$40x = 160$

$x = 4$

AQA Examiner's tip

You can try values to see whether they work or you can set up and solve an equation. Setting up and solving an equation is the most efficient method and is likely to take less time.

The numbers on the cards are

Card **A** $7 \times 4 = 28$ Card **B** $3 \times 4 = 12$ Card **C** $7 \times 4 - 10 = 18$ Card **D** $3 \times 4 + 30 = 42$

The ratio of the numbers on cards **A** and **D** = 28 : 42. This simplifies to $28 \div 14 : 42 \div 14 = 2 : 3$

The ratio of the numbers on cards **B** and **C** = 12 : 18. This simplifies to $12 \div 6 : 18 \div 6 = 2 : 3$

So the ratio of the numbers on cards **A** and **D** is the same as the ratio of the numbers on cards **B** and **C**.

AQA Examiner's tip

Always check the answers to problems. For example, in this question:

Card **A** = 28, Card **B** = 12, Card **C** = 18, Card **D** = 42

The ratio of numbers on cards **A** and **B** is 28 : 12 = 7 : 3

The ratio of numbers on cards **C** and **D** is 18 : 42 = 3 : 7

The numbers are in the ratios given in the problem.

Example

Unit 2

B

n is an integer.

Show that $2(3n + 4) + 4(n + 3)$ is **always** a multiple of 10.

Aim higher

You will need to be able to answer questions where you have to decide what you need to do to get a Grade A or A*.

Solution

For an algebraic expression to be a multiple of 10, it **must** have a factor of 10.

The first step is to simplify $2(3n + 4) + 4(n + 3)$

$$2(3n + 4) + 4(n + 3) = 6n + 8 + 4n + 12$$
$$= 10n + 20$$
$$= 10(n + 2)$$

So $2(3n + 4) + 4(n + 3)$ has a factor of 10 and must, therefore, be a multiple of 10.

Example

Unit 3

A*

O, A, B and C are four points such that

$\overrightarrow{OA} = 2\mathbf{a} - 4\mathbf{b}$, $\overrightarrow{OC} = 8\mathbf{a} - 3\mathbf{b}$ and $\overrightarrow{BC} = 2\mathbf{a} + 9\mathbf{b}$

Show that O, A and B lie on a straight line.

C

B

A

O

Not drawn accurately

Solution

You are given $\overrightarrow{OA}$. So to show that O, A and B lie on a straight line you need to work out $\overrightarrow{AB}$.

$\overrightarrow{AB} = \overrightarrow{AC} + \overrightarrow{CB}$

$\overrightarrow{AC} = \overrightarrow{AO} + \overrightarrow{OC}$

$\overrightarrow{AB} = \overrightarrow{AO} + \overrightarrow{OC} + \overrightarrow{CB}$

$\overrightarrow{AB} = -\overrightarrow{OA} + \overrightarrow{OC} + -\overrightarrow{BC}$

$\overrightarrow{AB} = -2\mathbf{a} + 4\mathbf{b} + 8\mathbf{a} - 3\mathbf{b} - 2\mathbf{a} - 9\mathbf{b}$

$\overrightarrow{AB} = 4\mathbf{a} - 8\mathbf{b} = 2(2\mathbf{a} - 4\mathbf{b}) = 2\overrightarrow{OA}$

OA and AB are parallel and have the point A in common, which means they must lie on a straight line.

Practise... Problem-solving

D C A

1 Chandi has four different coins in her pocket.

She picks three of the coins at random.

It is **not** possible for the total amount to be more than 72 pence.

Write a list of all the possible total amounts of money Chandi can take from her pocket.

You **must** show your working.

2 Show that $ABCD$ is a parallelogram.

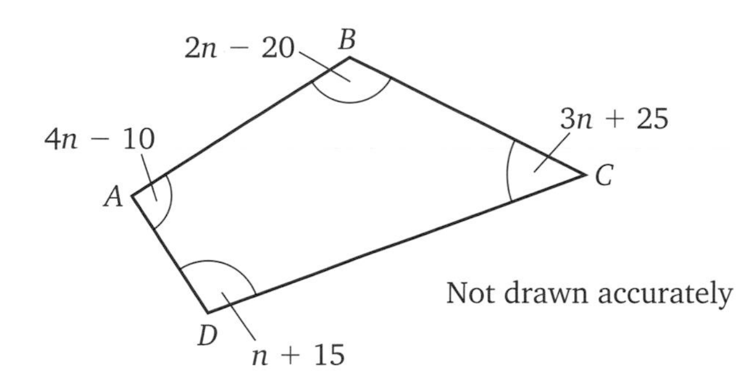

Not drawn accurately

3 At a gym, a notice on the rowing machine says '*On this machine you use 6 calories a minute*'.

On the cross-trainer it says '*On this machine you use 9 calories a minute*'.

Tom trains twice as long on the cross-trainer as he does on the rowing machine.

Altogether he uses 300 calories.

How long does he spend on each machine?

4 Bill thinks of a number.

He subtracts 3 from his number and then multiplies by 4.

Sally thinks of the same number as Bill.

She multiplies the number by 2 and then adds 15.

Bill and Sally get the same answer.

What number do they both think of?

5 It takes a gardener $2\frac{1}{2}$ minutes to tidy up one metre of the edge of flower beds.

How long does it take the gardener to tidy up a circular flower bed of diameter 3.2 metres?

Give your answer to the nearest minute.

6 The perimeter of rectangle **A** is 49 cm.

The perimeter of rectangle **B** is 36 cm.

Which rectangle has the greatest area?

Show how you decide.

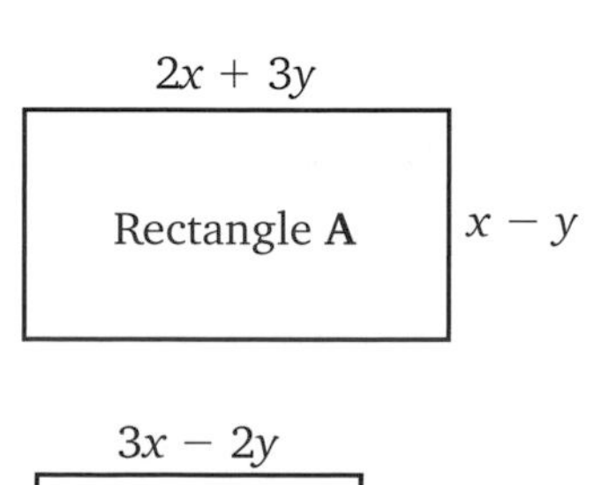

Essential skills

2 Proof

Units 2 3

Number and algebra
Geometry and algebra

Revise... Key points

A mathematical proof is a logical step-by-step procedure that shows how a statement or a proposition is true beyond any doubt.

Each step of the procedure must be mathematically valid and fully explained.

Many geometrical proofs were written by Euclid, a mathematician who lived around 300 BC. In fact, a lot of the geometry you need to know for Unit 3 of your GCSE exam is contained in his book *The Elements*!

Can you see how Euclid used this diagram to prove

'Of every triangle, when one of the sides is extended the external angle is equal to the two interior and opposite angles and the three interior angles of the triangle are equal to two right angles.'
(Euclid's proposition 32)

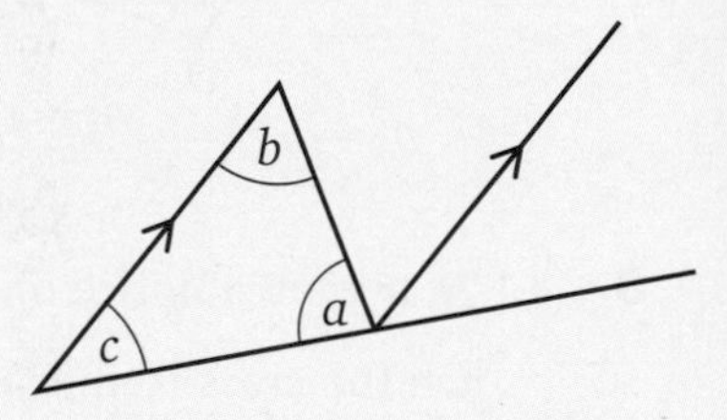

In proving the proposition, Euclid made use of these self-evident truths (things that are obviously true):

1 the sum of the angles on a straight line equals two right angles

2 the properties of parallel lines.

Self-evident truths like this form the basis of the geometrical proofs you might be asked for in Unit 3.

You could also be asked for algebraic proofs in Unit 2.

Aim higher

You will need to be able to answer questions asking you to give a simple proof to get a Grade A or A*.

Key terms

These terms will be used in your exam. Make sure you understand what they mean, and write your own definitions.

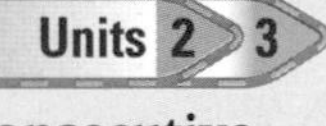

Consecutive
Proof
Prove

Example Algebra Unit 2

A

Tom is investigating the sum of three consecutive square numbers.

Here is some of his work.

$1^2 + 2^2 + 3^2 = 14 = 3 \times 4 + 2 = 3 \times 2^2 + 2$

$2^2 + 3^2 + 4^2 = 29 = 3 \times 9 + 2 = 3 \times 3^2 + 2$

$3^2 + 4^2 + 5^2 = 50 = 3 \times 16 + 2 = 3 \times 4^2 + 2$

$9^2 + 10^2 + 11^2 = 302 = 3 \times 100 + 2 = 3 \times 10^2 + 2$

Tom's work demonstrates that
$n^2 + (n + 1)^2 + (n + 2)^2 \equiv 3(n + 1)^2 + 2$

Prove this result.

The identity symbol, $\equiv$, is used because $n^2 + (n + 1)^2 + (n + 2)^2 \equiv 3(n + 1)^2 + 2$ is true for all values of n.

Solution

AQA Examiner's tip

To prove the identity you must

Start with $n^2 + (n + 1)^2 + (n + 2)^2$

Then use a step-by-step approach to prove it is the same as

$$3(n + 1)^2 + 2$$

It is a good idea to explain each step as you go along.

Expand the brackets $n^2 + (n + 1)^2 + (n + 2)^2 \equiv n^2 + (n^2 + 2n + 1) + (n^2 + 4n + 4)$

Collect like terms $n^2 + (n + 1)^2 + (n + 2)^2 \equiv 3n^2 + 6n + 5$

The result you are asked to prove gives you a clue about what to do next.
The expression on the right-hand side needs to be written with + 2 at the end.

$$n^2 + (n + 1)^2 + (n + 2)^2 \equiv 3n^2 + 6n + 3 + 2$$

$$n^2 + (n + 1)^2 + (n + 2)^2 \equiv 3(n^2 + 2n + 1) + 2$$

Factorise $n^2 + 2n + 1$

$$n^2 + (n + 1)^2 + (n + 2)^2 \equiv 3(n + 1)^2 + 2$$

Tom's work only **demonstrates** the result for the particular cases he tries.

The algebra proves that the result is true for **all** values of n.

AQA Examiner's tip

You will be assessed on the quality of your written communications, so it is important to set out your work in an organised way and write a conclusion. This should be a short statement like this.

Example Algebra Unit 2

A*

Prove that $\dfrac{x + y}{2(x - y)} + \dfrac{x - y}{2(x + y)} \equiv \dfrac{x^2 + y^2}{x^2 - y^2}$

Aim higher

You need to know how to manipulate algebraic fractions to get a Grade A or A*.

Solution

AQA Examiner's tip

To prove the identity you must start with

$$\frac{x + y}{2(x - y)} + \frac{x - y}{2(x + y)}$$

Then use a step-by-step approach to prove it is the same as

$$\frac{x^2 + y^2}{x^2 - y^2}$$

Explain each step as you go along.

Fractions can only be added if they have the same denominator.

The common denominator of $\dfrac{x + y}{2(x - y)}$ and $\dfrac{x - y}{2(x + y)}$ is $2(x - y)(x + y)$. This simplifies to $2(x^2 - y^2)$

So rewrite each fraction as an equivalent fraction with dominator $2(x^2 - y^2)$

$$\frac{x + y}{2(x - y)} \equiv \frac{(x + y)(x + y)}{2(x - y)(x + y)} \equiv \frac{(x + y)^2}{2(x^2 - y^2)} \qquad \frac{x - y}{2(x + y)} \equiv \frac{(x - y)(x - y)}{2(x - y)(x + y)} \equiv \frac{(x - y)^2}{2(x^2 - y^2)}$$

So $\dfrac{x + y}{2(x - y)} + \dfrac{x - y}{2(x + y)} \equiv \dfrac{(x + y)^2}{2(x^2 - y^2)} + \dfrac{(x - y)^2}{2(x^2 - y^2)} \equiv \dfrac{(x + y)^2 + (x - y)^2}{2(x^2 - y^2)}$

Now expand the brackets in the numerator.

$$\frac{(x+y)^2+(x-y)^2}{2(x^2-y^2)} \equiv \frac{(x^2+2xy+y^2)+(x^2-2xy+y^2)}{2(x^2-y^2)} \equiv \frac{2x^2+2y^2}{2(x^2-y^2)}$$

$$\frac{2x^2+2y^2}{2(x^2-y^2)} \equiv \frac{2(x^2+y^2)}{2(x^2-y^2)}$$

Dividing both the numerator and denominator by 2 gives $\frac{x^2+y^2}{x^2-y^2}$

So $\quad \frac{x+y}{2(x-y)} + \frac{x-y}{2(x+y)} \equiv \frac{x^2+y^2}{x^2-y^2}$

Example Geometry Unit 3

D A

a is Grade D, **b** is Grade A.

The diagram shows a quadrilateral with interior angles a, b, c and d and exterior angles w, x, y and z.

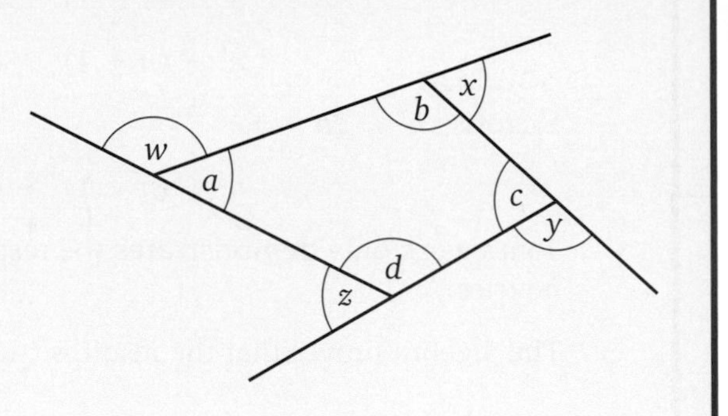

a **i** Explain why $a + w = b + x = c + y = d + z$

ii Explain why $w + x + y + z = 360°$

b Use your answers to part **a** to prove that the sum of the interior angles of a quadrilateral is 360°.

AQA Examiner's tip

Part **a** of this question is called a lead-in question. It is included to remind you of the maths to use in part **b**.

Solution

a **i** The angles are pairs of interior and exterior angles and lie on a straight line.

So $\quad a + w = 180°, b + x = 180°, c + y = 180°$ and $d + z = 180°$

ii Moving clockwise from the mid-point of any side along the sides of the quadrilateral, at each vertex you turn through an exterior angle.

So going all the way around the polygon back to the starting point you turn through angles w, x, y and z **and** a complete turn.

So, $w + x + y + z = 360°$

b $a + w = 180°, b + x = 180°, c + y = 180°$ and $d + z = 180°$

So $a + w + b + x + c + y + d + z = 4 \times 180$

So $(a + b + c + d) + (w + x + y + z) = 4 \times 180$

$w + x + y + z = 360°$

So $(a + b + c + d) + 360 = 4 \times 180$

So $a + b + c + d = 4 \times 180 - 360 = 360$

So the sum of the interior angles of a quadrilateral is 360°.

AQA Examiner's tip

This is a good example of how simple algebra and basic geometry can be combined to prove a geometrical fact.

Example Geometry Unit 3

A

The diagram shows a circle with centre O.

AB is a chord.

C is the mid-point of AB.

a Prove that triangle OAC is congruent with triangle OBC.

b Hence, prove OC is perpendicular to AB.

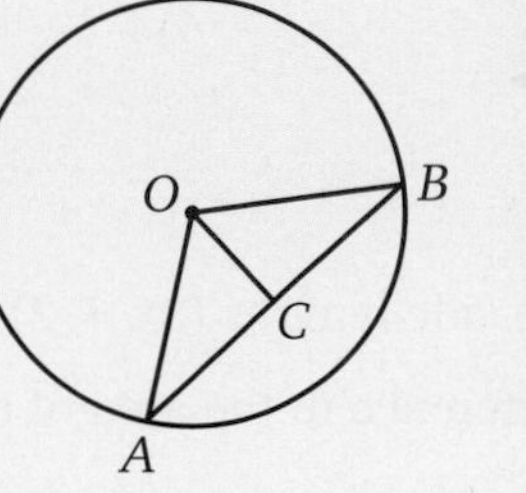

Link

Look back at Unit 3 Chapter 15 of your Student Book to remind yourself of the conditions for triangles to be congruent.

AQA Examiner's tip

To prove two triangles are congruent:

- Draw separate diagrams of each triangle.
- Mark equal lengths and equal angles on the triangles.
- Make one statement at a time and give a reason why it is true.
- State which of the conditions of congruency (SSS, SAS, ASA, RHS) is satisfied.

Solution

a

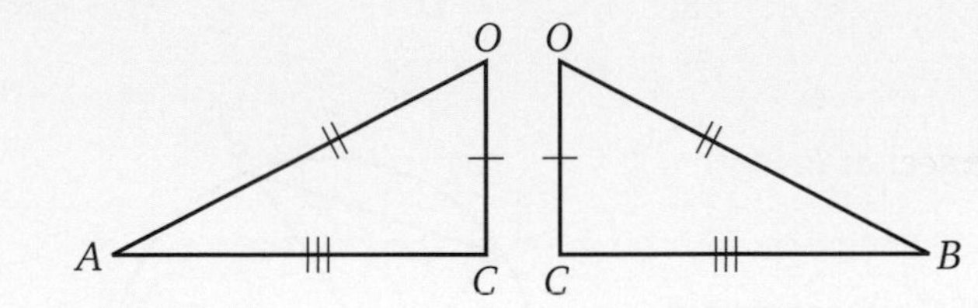

$OC = OC$ OC is a common length in both triangles.

$OA = OB$ OA and OB are both radii of the circle.

$AC = CB$ It is given that C is the midpoint of AB.

So triangle OAC is congruent to triangle OBC (SSS).

b Triangles OAC and OBC are congruent so angle OCA = angle OCB

Angle OCA + angle $OCB = 180°$ (The angles are on a straight line.)

So angle OCA = angle $OCB = 90°$

So OC is perpendicular to AB.

Practise... Proof

A

1 Prove that the sum of five consecutive integers is a multiple of 5.

2 **a** n is an integer.
Tim says $2n + 1$ must be an odd number.
Is Tim correct?
Show how you decide.

b Prove that the sum of three consecutive odd numbers is a multiple of 3.

3 Prove that the product of two consecutive odd numbers is 1 less than a multiple of 4.

4 A rectangle has length $\frac{x-1}{x+1}$ and width $\frac{x-3}{x^2-1}$

Prove that the perimeter of the rectangle is $\frac{2(x-2)}{x+1}$

A

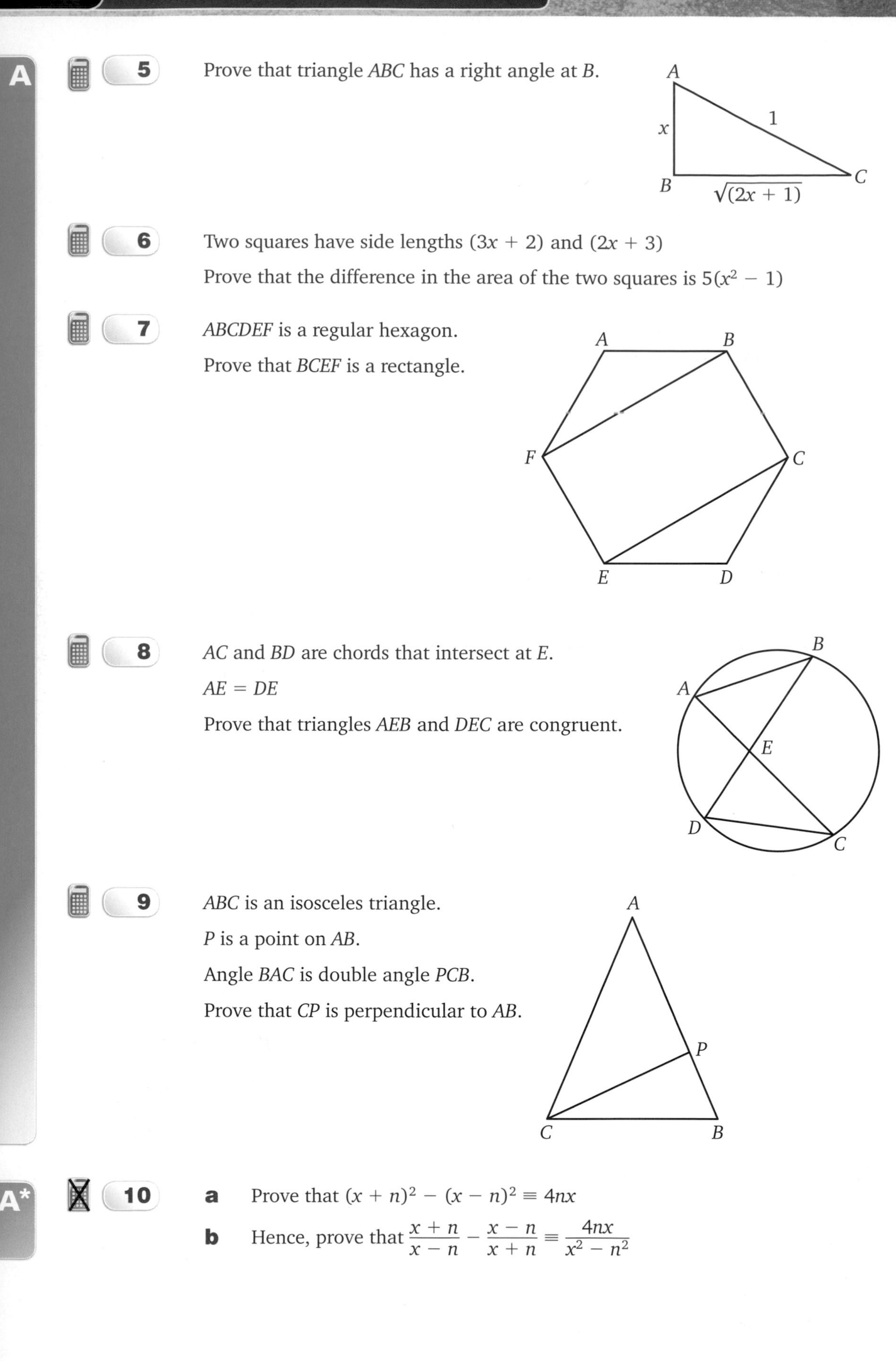

5 Prove that triangle ABC has a right angle at B.

6 Two squares have side lengths $(3x + 2)$ and $(2x + 3)$

Prove that the difference in the area of the two squares is $5(x^2 - 1)$

7 $ABCDEF$ is a regular hexagon.

Prove that $BCEF$ is a rectangle.

8 AC and BD are chords that intersect at E.

$AE = DE$

Prove that triangles AEB and DEC are congruent.

9 ABC is an isosceles triangle.

P is a point on AB.

Angle BAC is double angle PCB.

Prove that CP is perpendicular to AB.

A*

10

a Prove that $(x + n)^2 - (x - n)^2 \equiv 4nx$

b Hence, prove that $\dfrac{x + n}{x - n} - \dfrac{x - n}{x + n} \equiv \dfrac{4nx}{x^2 - n^2}$

AQA Examination-style questions

AQA Examination-style questions

1 Tom makes a cuboid using 15 one-centimetre cubes.
Jerry makes a different cuboid using 15 one-centimetre cubes.
Tom's cuboid has a smaller surface area than Jerry's cuboid.
What are the surface areas of Tom and Jerry's cuboids?
You **must** show your working. *(5 marks)*

D

2 Donna tries to solve puzzles at different levels of difficulty from level 1 (easy) to level 10 (hard).
The table shows the times she takes to solve **some** of the puzzles.

Level of difficulty	1	2	3	4	5	6	7	8	9	10
Time (minutes)	6	11	13	18	25	24		38	45	53

a Donna is interested in the relationship between the level of difficulty and the time she takes to solve the puzzles.
On a suitable grid, draw a diagram to show the relationship.

b Estimate the time it takes Donna to solve a level 7 puzzle. *(3 marks)*

C

3 Each term of a Fibonacci sequence is formed by adding the previous two terms.

1, 1, 2, 3, 5, 8, 13, 21, ...

A Fibonacci sequence starts a, b, $a + b$, ...
Prove that the difference between the 9th term and the 3rd term of this sequence is four times the 6th term. *(5 marks)*

A

4 The diagrams show two different cans of baked beans.
They are made of metal of the same thickness.
The diameter of can **A** is 74 mm.
The height of can **A** is 108 mm.
The diameter, D, of can **B** equals its height, h.

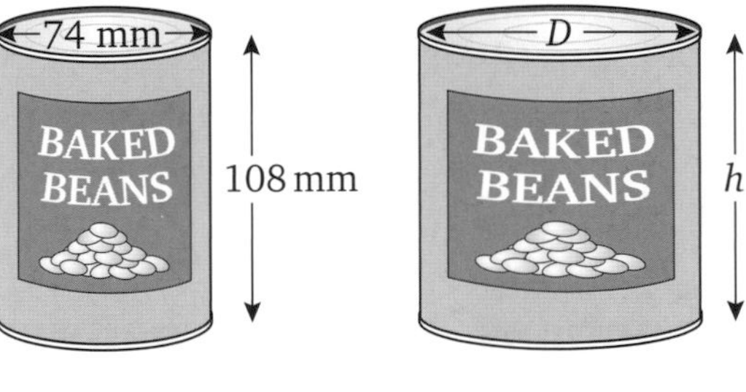

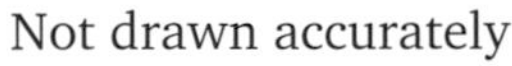
Not drawn accurately

a Can **A** and can **B** have equal volume.
Work out the dimensions of can **B** to the nearest millimetre.

b Show that can **B** is made from approximately $\frac{2}{3}$ of the amount of metal used to make can **A**. *(8 marks)*

A*

5 XYZ is an isosceles triangle in which $XZ = XY$
M and N are points on XZ and XY such that angle MYZ = angle NZY
Prove that triangles YMZ and ZNY are congruent.

X, N, M, Y, Z

(4 marks)

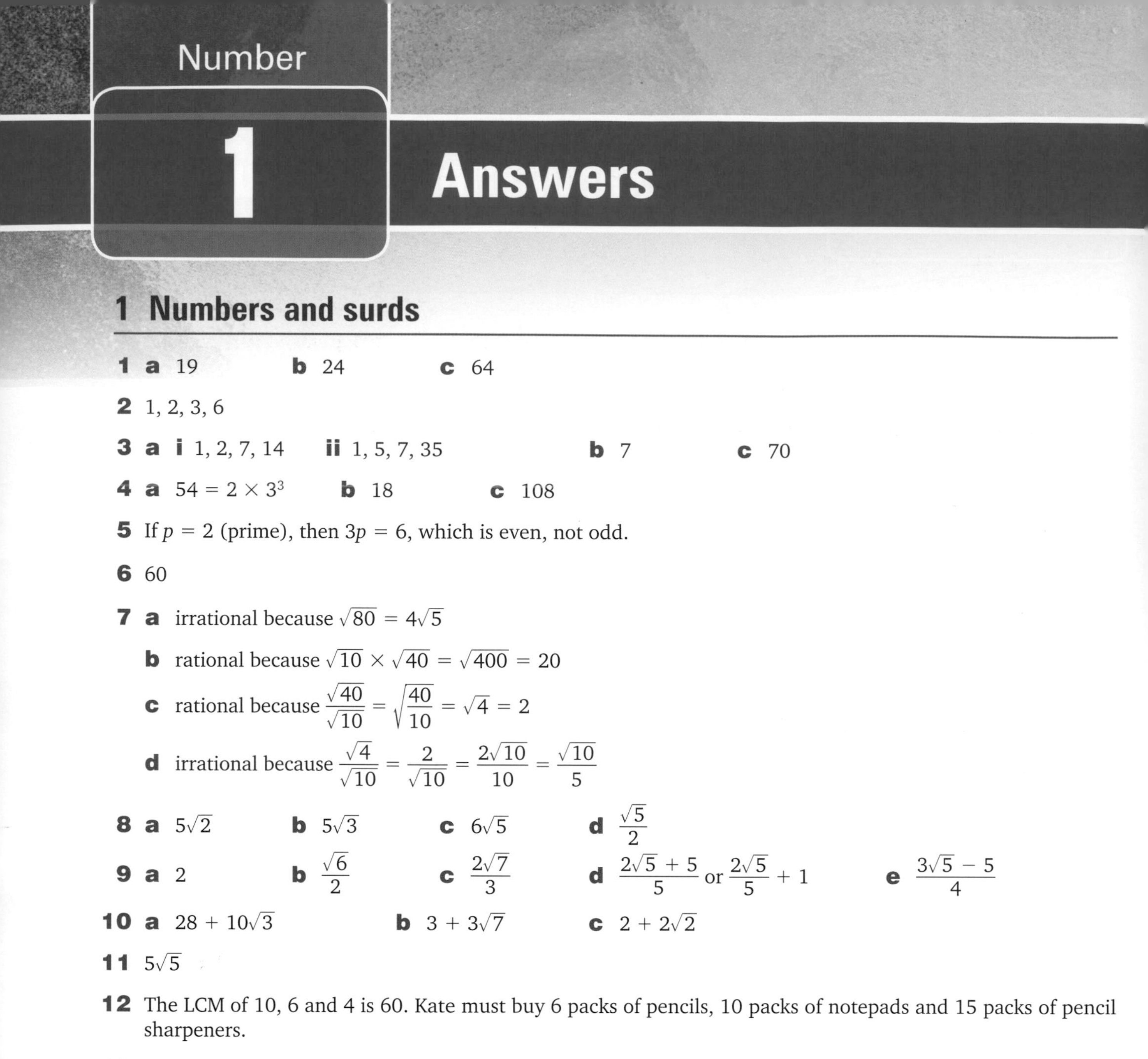

Number

1 Answers

1 Numbers and surds

1 a 19 **b** 24 **c** 64

2 1, 2, 3, 6

3 a i 1, 2, 7, 14 **ii** 1, 5, 7, 35 **b** 7 **c** 70

4 a $54 = 2 \times 3^3$ **b** 18 **c** 108

5 If $p = 2$ (prime), then $3p = 6$, which is even, not odd.

6 60

7 a irrational because $\sqrt{80} = 4\sqrt{5}$

b rational because $\sqrt{10} \times \sqrt{40} = \sqrt{400} = 20$

c rational because $\frac{\sqrt{40}}{\sqrt{10}} = \sqrt{\frac{40}{10}} = \sqrt{4} = 2$

d irrational because $\frac{\sqrt{4}}{\sqrt{10}} = \frac{2}{\sqrt{10}} = \frac{2\sqrt{10}}{10} = \frac{\sqrt{10}}{5}$

8 a $5\sqrt{2}$ **b** $5\sqrt{3}$ **c** $6\sqrt{5}$ **d** $\frac{\sqrt{5}}{2}$

9 a 2 **b** $\frac{\sqrt{6}}{2}$ **c** $\frac{2\sqrt{7}}{3}$ **d** $\frac{2\sqrt{5} + 5}{5}$ or $\frac{2\sqrt{5}}{5} + 1$ **e** $\frac{3\sqrt{5} - 5}{4}$

10 a $28 + 10\sqrt{3}$ **b** $3 + 3\sqrt{7}$ **c** $2 + 2\sqrt{2}$

11 $5\sqrt{5}$

12 The LCM of 10, 6 and 4 is 60. Kate must buy 6 packs of pencils, 10 packs of notepads and 15 packs of pencil sharpeners.

13 59 and 61 (difference = 2)

14 $3\sqrt{2}$

15 Yes. $(2 + \sqrt{3})^2 = 4 + 4\sqrt{3} + 3 = 7 + 4\sqrt{3}$ and $(\sqrt{2} + \sqrt{6})^2 = 2 + 2\sqrt{12} + 6 = 8 + 4\sqrt{3}$
So $1^2 + (2 + \sqrt{3})^2 = (\sqrt{2} + \sqrt{6})^2$ and the triangle has a right angle (Pythagoras).

2 Fractions, decimals and rounding

1 a 3.847926267 **b i** 3.8 **ii** 4

2 a 0.65 **b** 0.875 **c** $0.\dot{6}$ **d** $0.\dot{7}$ **e** $1.\dot{0}\dot{9}$

3 a $\frac{2}{5}$ **b** $\frac{3}{5}$

4 $\frac{7}{10}$

5 a $1\frac{13}{18}$ **b** $\frac{9}{40}$ **c** $\frac{9}{14}$ **d** $1\frac{1}{2}$

6 a $\frac{17}{20}$ **b** $\frac{14}{25}$ **c** $\frac{3}{8}$ **d** $\frac{2}{125}$ **e** $\frac{1}{16}$

7 a i 0.72 **ii** 12.6

8 a 300 **b** 2000

9 b, **e** and **g**

10 a $7\frac{1}{10}$ **b** $3\frac{5}{6}$ **c** $8\frac{4}{5}$ **d** $\frac{3}{4}$ **e** $1\frac{1}{2}$

11 Yes. $1\frac{1}{3} \times \frac{2}{3} = \frac{4}{3} \times \frac{2}{3} = \frac{8}{9}$ is less than 1 (or example with assumed price).

12 a $0.\dot{6}\dot{3}$ **b** $\frac{7}{9}$ and $\frac{1}{7}$ with working or explanation

13 a 211 mm **b** 59 mm

14 a $\frac{8}{9}$ **b** $\frac{5}{11}$ **c** $\frac{1}{22}$ **d** $\frac{4}{15}$ **e** $\frac{8}{37}$

15 5 tins **16** £11.30 **17** 4 hours **18** 4 pint + 6 pint **19** 285 grams **20** 26

21 No, Kieran is wrong. $\frac{1}{10} = \frac{6}{60}$ and $\frac{1}{20} = \frac{3}{60}$, but $\frac{1}{15} = \frac{4}{60}$ or $\frac{1}{10} = 0.1$ and $\frac{1}{20} = 0.05$, but $\frac{1}{15} = 0.0\dot{6}$

22 Any fraction between $\frac{3}{4}$ and $\frac{4}{5}$ that is a recurring decimal (for example $\frac{7}{9}, \frac{34}{45}, \frac{23}{30}, \frac{71}{90}$)

3 Percentages

1 a i £143.10 **ii** £126.90

2 a 24 **b** 20%

3 20%

4 25%

5 Jack's (4%, more than Jill's 3.5%)

6 Shop B is cheapest (Shop A price = £450, Shop B price = £448, Shop C price = £470)

7 $8\frac{1}{3}\%$ **8** £160 **9** 12%

10 a £7908.24 (nearest pence)
b After 11 years, the amount = £6500 × 1.04^{11} = £10 006.45…

11 a £18 740 (nearest £)
b After six years, the value = £27 500 × 0.88^{6} = £12 771.11…
c

27 500
Value (£)
0
0 6
Time (years)

4 Ratio and proportion

1 a $\frac{5}{9}$ **b** $\frac{4}{9}$ **c** 15 girls, 12 boys

2 54° **3** 65% **4** 3 : 80 : 20

5 a 50 people **b** 96°

6 8 days

7 a $P = 2.5Q^3$ **b** 4

8 a $y \propto \frac{1}{x^2}$ so $y = \frac{79.2}{x^2}$ **b**

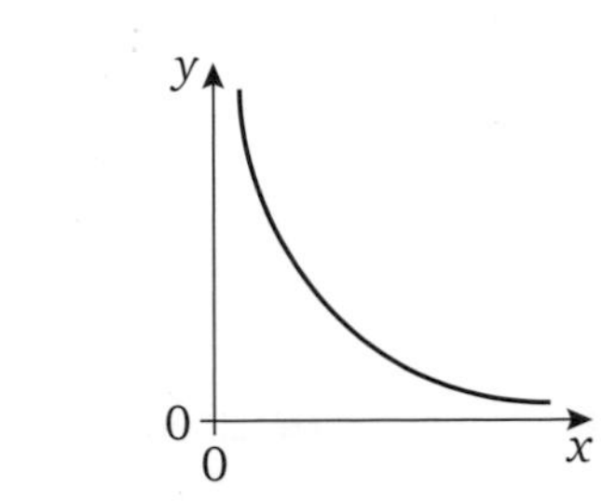

9 a B **b** A **c** D

10 Will gets £16, Kate gets £20 and Neil gets £28.

11 8.5 km

12 a School A 1 : 14.8 School B 1 : 14.2 (to 3 s.f.)
b School B because it is likely to have smaller class sizes

13 a Large bag gives the best value for money (Small 0.18p/g, Medium 0.161p/g, Large 0.1595p/g)
b She may not have enough room or money for the large bag.

14 Adel gets 12%, Barry gets 18%, Cathy gets 30%.

15 $p \propto \frac{1}{r}$

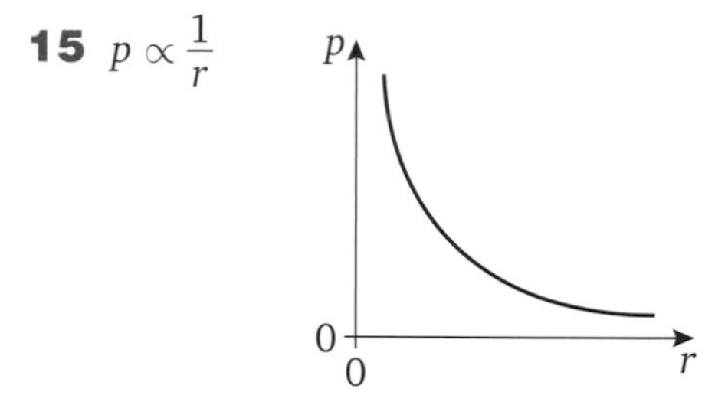

5 Indices and standard index form

1 a 0.9^2 **b** $\sqrt{0.36}$ **c** $\sqrt[3]{-27}$

2 a 48 **b** 10 **c** 120 **d** -1 **e** ± 9 **f** -2

3 a 10 000 **b** 10 000 **c** 32 **d** -32 **e** 625 **f** 1024

4 a 3^9 **b** 7^4 **c** 6^6 **d** p^{10} **e** q **f** r^{15} **g** 5^2 **h** 4^{21} **i** 8^{24} **j** 9^4 **k** t **l** n^6

5 a 144 **b** 40 **c** 9 **d** 2 **e** 1250

6 a Chloe has multiplied the indices instead of adding them. Dave has multiplied the base numbers.
b 3^7 or 2187

7 a i 8.56×10^6 **ii** 7.5×10^{-5} **b i** 260 000 000 **ii** 0.00624

8 9.3×10^{-4}, 1.82×10^{-3}, 5.7×10^{-2}, 6.8×10^0, 2.4×10^6, 1.95×10^7

9 a 6×10^5 **b** 2×10^{-3} **c** 1.4×10^5 **d** 1.5×10^8 **e** 6×10^{-2} **f** 1.8×10^{20}

10 a 7^{-1} **b** 2^8 **c** 5^{-3} **d** p^{-5} **e** b^{-6} **f** n^6 **g** 10^6 **h** y^{-9} **i** $10a^5$ **j** 1 **k** $8r^{-3}$ **l** $2p^2$

11 a 3 **b** 5 **c** 100 **d** $\frac{1}{4}$ **e** 1 **f** $\frac{1}{8}$

12 £360–£400

13 a Io, Europa, Ganymede, Callisto **b i** 2 650 000 km **ii** 2.7×10^6 km

14 a i £8.8×10^8 **ii** 4.4×10^9 **b** 20 pence

15 a 5×10^7 **b** 1.2%

16 a 7.8×10^9 euros **b** Music 41%, Games 36%, Video 23%

17 a The unit digit will always come from $1 \times 1 = 1$ **b** 0, 5 and 6

18 a i 4 **ii** 9 **iii** 16 **b** $100^2 = 10\,000$

19 a i 2^4 cm **ii** 2^2 cm^2 **iii** 2^8 cm^2 **iv** 2^7 cm^2 **b** 2^{63}

6 AQA Examination-style questions

1 **a** £18 952 **b** 3.2%

2 £1400

3 Shop Q (£110)

4 2 + any other prime number

5 1, 2, 7, 14

6 **a** **i** 0.16 **ii** 8 **b** **i** 18 000 **ii** 8 or 10

7 **a** ± 68 **b** 4.913 **c** **i** 12.88194444 **ii** 12.9 **d** $\frac{1}{64}$ or 0.015625

8 9 tins

9 **a** **i** $3\frac{1}{18}$ **ii** $2\frac{2}{3}$ **b** 4

10 **a** $2^3 \times 3^2$ **b** **i** $2^4 \times 3^2$ **ii** $2^4 \times 3^2 \times 5$

11 12 times

12 Offer A is better (0.252p/g is cheaper than 0.277p/g).

13 20%

14 60 g

15 **a** **i** t^{12} **ii** t^{-6} **iii** t^{27} **b** iii **c** ii

16 **a** $\frac{13}{20}$ **b** $\frac{4}{165}$

17 **a** £7804.67 **b** After 10 years, the amount = £6000 × 1.054^{10} = £10 152.13...

18 **a** 27.5 is greater than 10 **b** 3×10^6 **c** 0.0628

19 **a** 17 **b** **i** $6.31 \times 10^{22}\,m^3$ **ii** $1600\,kg/m^3$

20 **a** $y = \frac{500}{x^2}$ **b** y is divided by 4 **c**

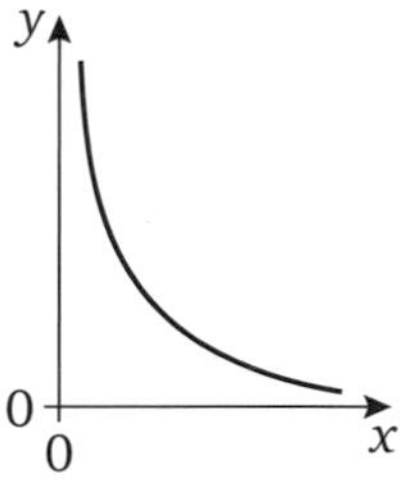

21 **a** $\frac{\sqrt{5}}{2}$ **b** $2\sqrt{3}$

22 **a** 9 **b** $-\frac{2}{3}$

23 **a** 150 million (to 3 s.f.)

b **i**

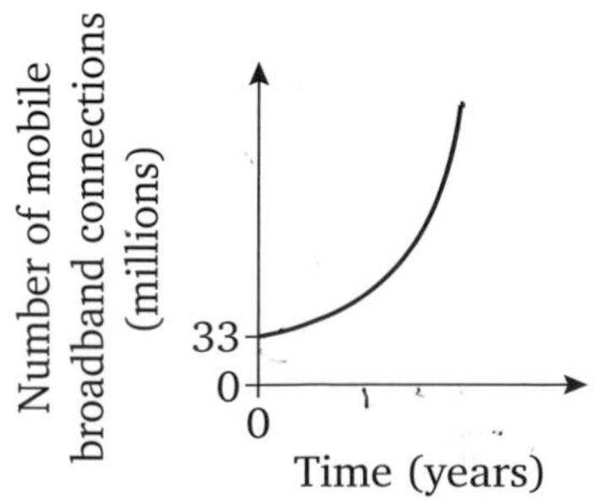

ii exponential growth

iii Growth will slow down as fewer people are left without a mobile broadband connection.

Statistics

2 Answers

1 Collecting data

1 **a** Method 1 – Unlikely to get a representative set of views, for example unlikely to get people who work or young people.

Method 2 – Not totally random as not everyone is in the phone book. Some people will therefore have no chance of being selected.

Method 3 – May not be representative if survey is only carried out in one part of town.

b List everyone in town and allocate each person a number. Select 200 numbers at random using a calculator or random number tables.

2 **a** It is a leading question/biased towards Pricewise.

b There is no 'Disagree' box.

c asking only customers in Pricewise, so biased

d This is primary data, as it was collected by someone standing outside the supermarket.

3 Student's own answer. Must include a description of how the data will be collected, an example of at least one calculation/diagram and how they will use this to make a conclusion.

4 Science: $\frac{80}{1000} \times 405 = 32$, English: 24, Mathematics: 15, Psychology: 9

5 **a** more representative, ensures views in the right proportions from all groups

b 37.9%

c Class A Boys: 3 Class A Girls: 4 Class B Boys: 6 Class B Girls: 2

d For each group put the students' names in a hat and randomly select the required number.

6 112 conductors

7 12.8, round to 13 boxes

2 Statistical measures

1 **a** The mean must lie between 3 and 6.

b 5.01

2 **a** 1 letter **c** 2 letters **e** $3 - 1 = 2$ letters

b 2 letters **d** $7 - 0 = 7$ letters **f** because 7 letters is an outlier

3 **a** 1.5 people per car (to 1 d.p.)

b 1.8 people per car (to 1 d.p.)

c It may have had some effect, as there are more people going into the city in fewer cars.

4 mean before campaign = 3.39 (3 s.f.)
mean after campaign = 3.62 (3 s.f.)

This suggests the advertising campaign has had some success, although on average people are still not eating 5 portions a day.

5 **a** They are using the modal or median class, which is $60 \leqslant x < 70$

b Estimate of mean is 70.75 mph.

6 26.9 marks

7 58.2%

3 Representing data

1 a 16 **b** 5 **c** 20.5 seconds **d** 39 seconds

2 a frequency polygon (points at (25, 5), (35, 11), (45, 24), (55, 36), (65, 8))
b The first exercise class is more popular.
c One point comparing average, and one point comparing range. Generally younger people attend the second exercise class – the mode is lower down the graph. The range of ages is higher for the first exercise class – diagram is more spread out.

3 a Plot scatter graph. **b** 19–21 mg of salt **c** not reliable as trend may not continue

4 440 pies

5 histogram with frequency densities (4, 8.5, 7, 5.2, 1.5)

6 a 75 flowers **b** 40 cm

7 47.6 marks (accept 47 or 48 marks)

4 Probability

1 a 0.7 **b** 0.1 **c** 0.7 **d** 36

2 27 games

3 a $\frac{3}{10}$ or 0.3
b $\frac{11}{40}$ or 0.275
c Plot relative frequency polygon.
d $\frac{21}{80} = 0.2625$ (allow between 0.25 and 0.27)

4 a $\frac{42}{100}$ or 0.42 **b** $\frac{42}{90}$ or 0.46…

5 a $\frac{28}{90}$ **b** $\frac{34}{90}$ **c** $\frac{70}{90}$

6 0.41

7 Probability of winning is greater than $\frac{1}{3}$ (0.371 to 3 d.p.), so suggested amount to pay out is £1.50 to £2.00. Should not pay out more than £2.75 as charity would then be losing money.

8 $\frac{13}{36}$

5 AQA Examination-style questions

1 24 times

2 a 40 houses
b 4 bedrooms
c Median = 4 bedrooms Mean = 3.7 bedrooms (to 1 d.p.)

3 a

Stem	Leaf
0	8
1	2 5 6 7 9
2	2 5 5 6 7 8 8
3	2 3 5 6 9
4	0 1

Key: 1 | 2 represents 12 lengths

b Median before = 26.5 lengths Median after = 26 lengths
Therefore median has decreased.

4 a Scatter diagram plotted.

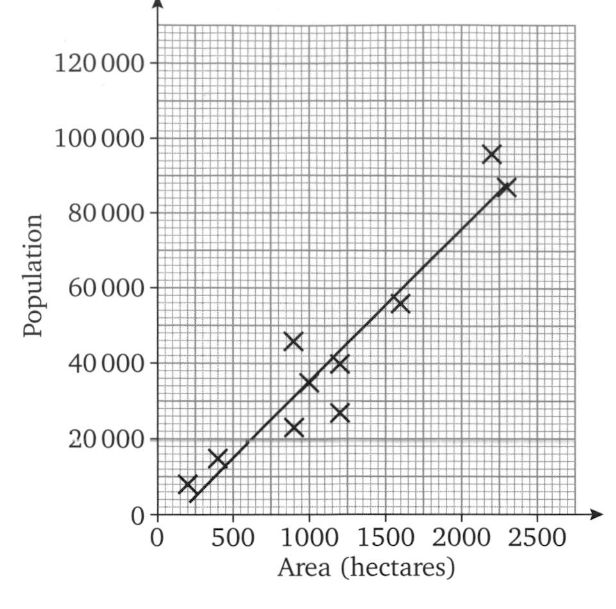

b quite strong positive correlation

c line of best fit drawn

d answer between 750 and 850 hectares.

5 No. Girls' mean is £5.70 and boys' mean is £6.93.

6 a 5 times

b

Score	1	2	3	4	5	6
Relative frequency	$\frac{3}{30}$	$\frac{5}{30}$	$\frac{14}{30}$	$\frac{3}{30}$	$\frac{3}{30}$	$\frac{2}{30}$

c The dice appears to be biased towards 3 because 3 is the result almost half the time.

7 Student's own answer. Must include a description of how the data will be collected, an example of at least one calculation/diagram, and how they will use them to make a conclusion.

8 a Cumulative frequency diagram drawn with following points:

(5, 0), (10, 5), (15, 17), (20, 31), (25, 38) and (30, 40)

b i Median = 16.1 **ii** Inter-quartile range = 6.7

c Box plot plotted with following information:
Lowest: 0 Lower quartile: 12.9 Median: 16.1 Upper quartile: 19.6 Largest: 30

d Generally trees in second wood have larger diameter (higher median).
Larger range of diameters in the second wood.

9 a $\frac{12}{49}$

b $\frac{37}{49}$

c New probability is $\frac{10}{42}$ and $\frac{10}{42}$ is less than $\frac{12}{49}$

10 answer in the region of 110 bpm

11 Bungalow 17, Terraced 63, Semi-detached 190, Detached 30
or Bungalow 18, Terraced 62, Semi-detached 190, Detached 30

12 a Bar from 4–6 goes to frequency density of 11.
Bar from 6–10 goes to frequency density of 3.

b Jake is correct.
Morning: 12 calls less than 1 minute. Evening: 14 calls less than 1 minute

13 a $\frac{29}{52}$ **b** $\frac{391}{780}$

3 Answers

1 Sequences and symbols

1 a 4, 9, 14 **b** 17th term is 84, which is 6×14
c All terms are one less than a multiple of 5 and 205 is a multiple of 5.

2 a $8k - 4$ **b** $6m - m^2$ **c** $6t^2 + 4t$

3 a $5(2x + y)$ **b** $t(t - 5)$ **c** $3k(k - 3)$

4 $(2p + 1)$ cm

5 a $3n - 1$ **b** $2.5n - 1.5$ or $\frac{5n - 3}{2}$

6 a $10a + 5b$ **b** $5x - 12y$

7 43, 124

8 i True **ii** False **iii** True **iv** False

9 a $x^2 + 11x + 24$ **b** $y^2 - 3y + 2$ **c** $z^2 - z - 12$ **d** $2x^2 - 9x - 5$ **e** $3y^2 + y - 10$ **f** $4z^2 + 19z - 30$ **g** $x^2 + 14x + 49$ **h** $4y^2 - 1$ **i** $9t^2 - 12t + 4$

10 a $\frac{8a}{15}$ **b** $\frac{11b}{24}$ **c** $\frac{2de}{5}$ **d** $\frac{p^2}{3}$ **e** $\frac{2}{3}$ **f** $\frac{3}{p - q}$

2 Equations and formulae

1 a $k = 5$ **b** $m = -2$ **c** $n = 0.5$ **d** $x = 2.8$

2 $3x + 5x + x + 50 + x - 10 = 360$, $x = 32$

3 £2142.45

4 A is an identity, B is an equation, C is a formula. Expression could be $3x - 4$ (free choice).

5 a $x = -6$ **b** $y = 1$ **c** $t = -10$

6 $3(2x + 15) = 69$, $x = 4$

7 a $x = -6$ **b** $y = 6.5$ **c** $z = 48$

8 a $b = \frac{P - 3a}{2}$ **b** $x = \sqrt{\frac{V}{h}}$

9 a $y = 13$ **b** $p = 7$

3 Trial and improvement

1 $x = 1.8$

2 $y = 2.4$

3 $t = 4.1$

4 a $m = -2.2$ **b** $m = 2.9$

5 $x = -4.4$

6 $p = 6.8$

7 $y = 3.7$

4 Coordinates and graphs

1 a/b

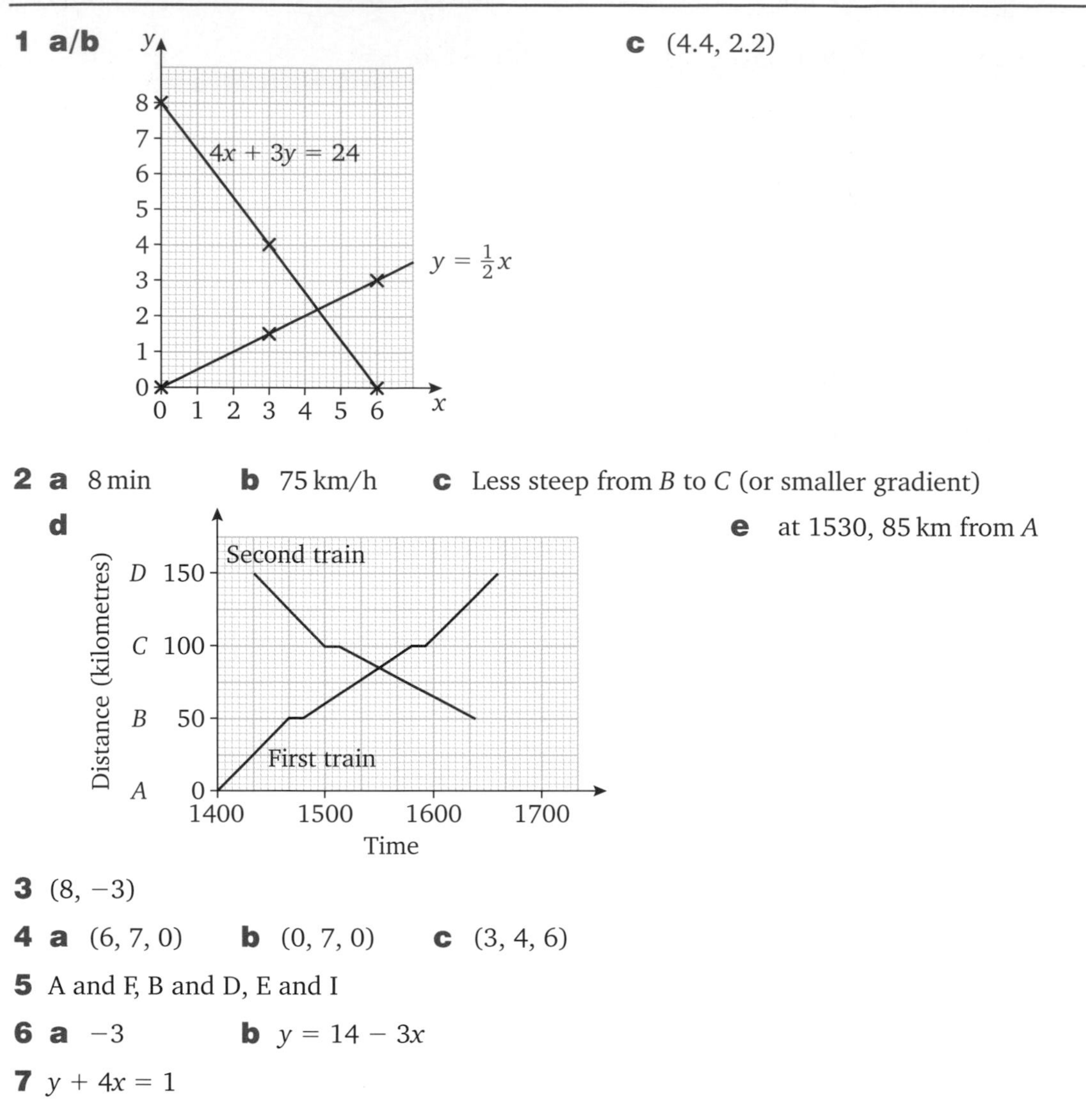

c (4.4, 2.2)

2 a 8 min **b** 75 km/h **c** Less steep from B to C (or smaller gradient)

d

e at 1530, 85 km from A

3 (8, −3)

4 a (6, 7, 0) **b** (0, 7, 0) **c** (3, 4, 6)

5 A and F, B and D, E and I

6 a −3 **b** $y = 14 - 3x$

7 $y + 4x = 1$

5 Quadratic functions

1 $12x - 8 + 2x - 20 = 14x - 28 = 14(x - 2)$

2 a

x	−3	−2	−1	0	1	2	3
y	−6	**−1**	2	3	**2**	−1	**−6**

b $x = 0$

b

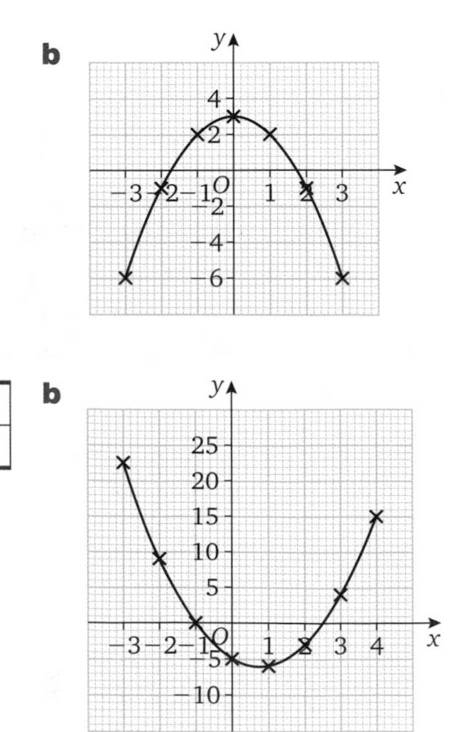

3 a

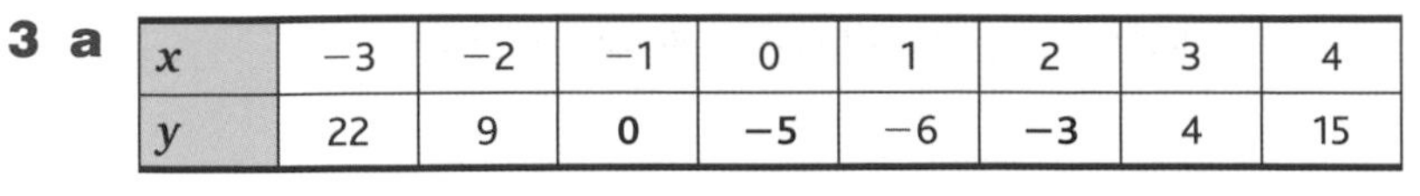

x	−3	−2	−1	0	1	2	3	4
y	22	9	**0**	**−5**	−6	**−3**	4	15

c $x = -1.4$ or 2.9

b

4 a $(p + 1)(p + 4)$ **b** $(q - 9)(q + 4)$ **c** $(6 + t)(5 - t)$ **d** $(3 - v)(3 + v)$ **e** $(2x + 3)(x + 2)$ **f** $(3y - 5)(y - 1)$ **g** $(5z - 4)(z + 3)$ **h** $(2w - 5)(2w + 5)$

5 a $3(x - 3)(x - 2)$ **b** $2(y - 2)(y + 2)$ **c** $t(t + 6)(t - 4)$

6 a $a = -5$ or 8 **b** $b = -7$ or $\frac{3}{4}$ **c** $c = 3$ or 4 **d** $d = -2$ or 1 **e** $e = -10$ or 3 **f** $f = -1$ or $-\frac{1}{2}$ **g** $g = -2\frac{2}{3}$ or 2 **h** $h = -2$ **i** $k = \pm\frac{1}{4}$ **j** $m = -\frac{1}{4}$ or 0

7 a $\frac{a + 3}{a - 3}$ **b** $\frac{b - 7}{b - 9}$ **c** $\frac{c - 10}{c + 5}$ **d** $\frac{5d - 2}{d - 2}$ **e** $\frac{e - 2}{3e + 2}$ **f** $\frac{3f - 1}{2f - 1}$

8 a $x = 1\frac{1}{2}$ or 4 **b** $y = -2$ or 19 **c** $t = -1\frac{2}{3}$ or 11

9 a $a = -5.37$ or 0.37 **b** $b = 1.44$ or 5.56 **c** $c = -2.77$ or 1.27 **d** $d = 0.28$ or 2.39 **e** $e = -1.41$ or 3.91 **f** $f = -1.36$ or 1.11

10 $b^2 - 4ac = 25 - 48 = -23$, which does not have a square root

11 Call first integer x.
Second integer is $x + 1$.
$(x + 1)^2 - x^2 = x^2 + 2x + 1 - x^2 = 2x + 1$
$2x$ is always even, so $2x + 1$ is always odd.

12 a $a = 11$ **b** $b = 56$ **c** $k = -6 \pm \sqrt{31}$

6 Higher graphs

1 a

x	−3	−2	−1	0	1	2	3
y	−13	**3**	7	**5**	3	**7**	23

b

c i $x = -2.3$ **ii** $x = -1.9, 0.4, 1.5$

2 a Base of box is a square of length $(20 - 2x)$ cm. Height of box is x cm.
V = area of base × height $= x(20 - 2x)^2$

b When $x = 0$ the box has no height.
When $x = 10$ there is nothing left after the squares are cut out.

c

x	0	1	2	3	4	5	6	7	8	9	10
V	0	**324**	512	588	**576**	**500**	384	252	**128**	**36**	0

d

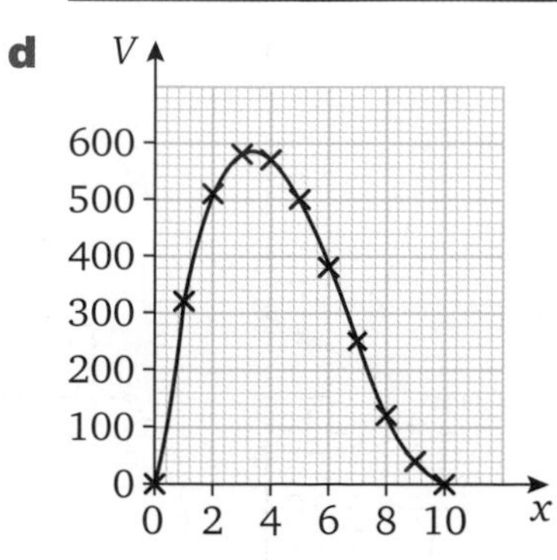

e Max V approx 593 cm^3, when $x = 3.3$

3 a $xy = 45$ or $y = \frac{45}{x}$

b

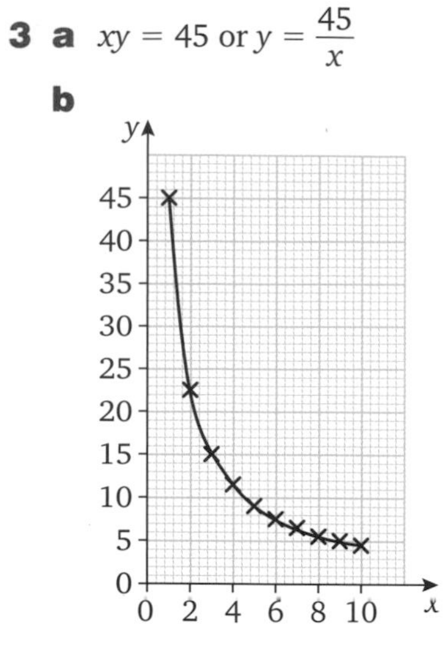

c Length = 5.6 cm

4 a

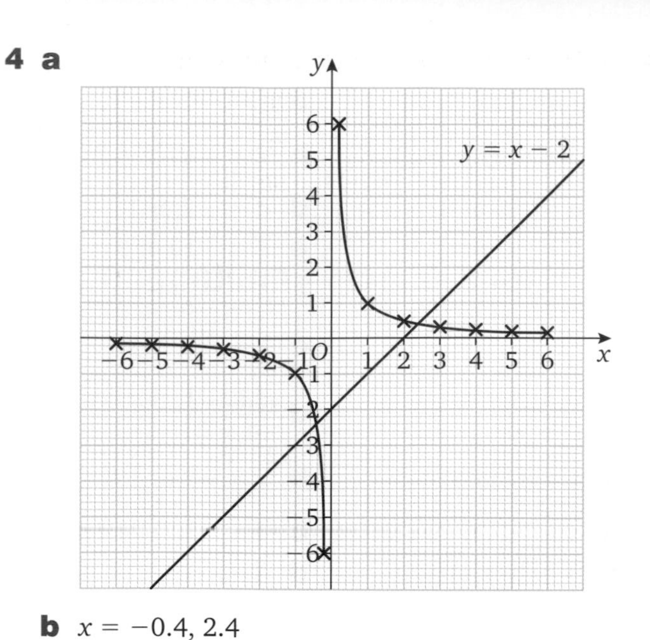

b $x = -0.4, 2.4$

5 a

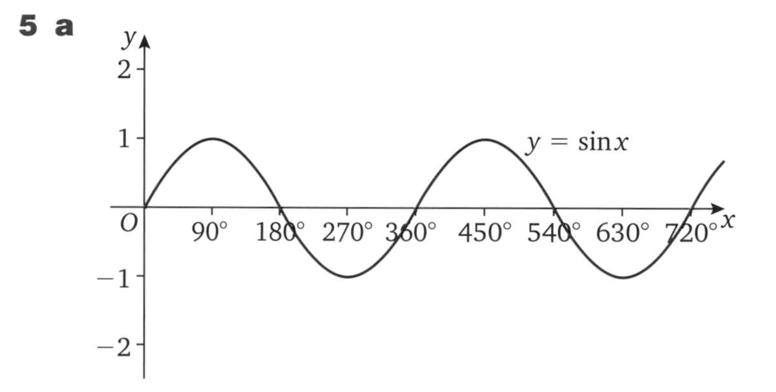

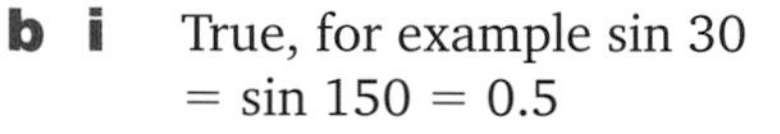

b i True, for example sin 30 = sin 150 = 0.5

ii False, for example sin 330 = −0.5 but sin 30 = 0.5

iii True, for example sin 390 = sin 30 = 0.5

iv True, for example sin 690 = sin 210 = −0.5

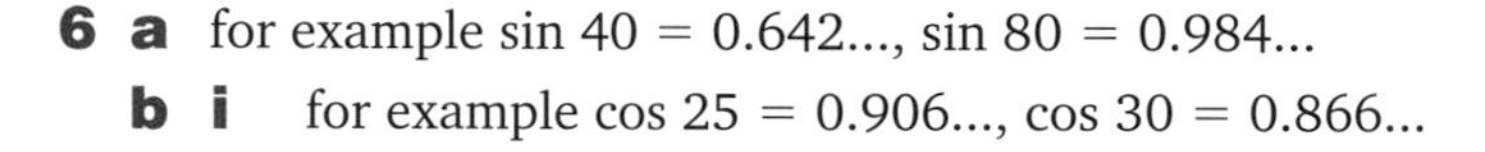

6 a for example sin 40 = 0.642..., sin 80 = 0.984...

b i for example cos 25 = 0.906..., cos 30 = 0.866...

ii

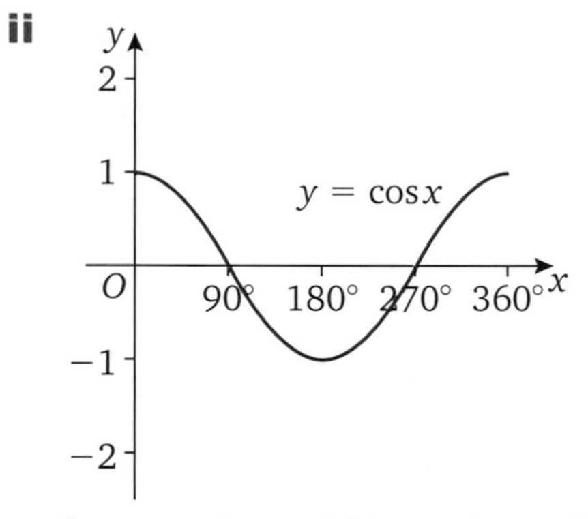

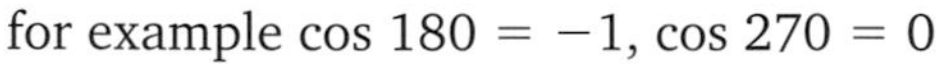

for example cos 180 = −1, cos 270 = 0

7 a 720

b

x	0	1	2	3	4	5	6	7	8	9	10
B	500	600	720	864	1037	1244	1493	1792	2150	2580	3096

c

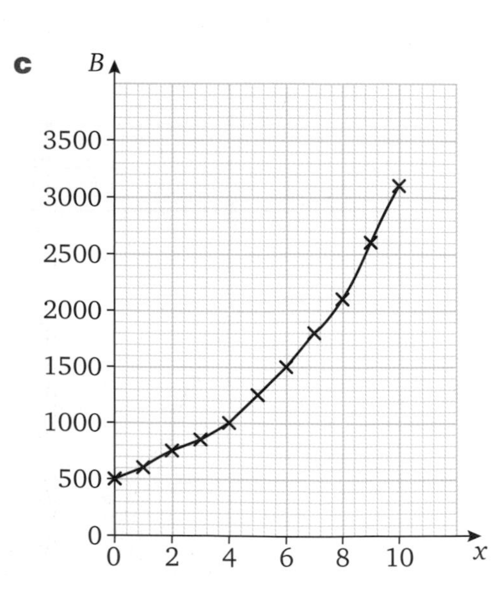

d after 6.4 hours

8 a Graph does not go through the origin.

b Graph is not symmetrical about the y-axis.

c $xy = 5$ would not touch the y-axis and when x is small, y would be very large.

d An exponential graph has small values of y when x is negative and larger ones when x is positive.

7 Inequalities and simultaneous equations

1 a $-1 < x$ **b** $-3 \leqslant x < 4$

2

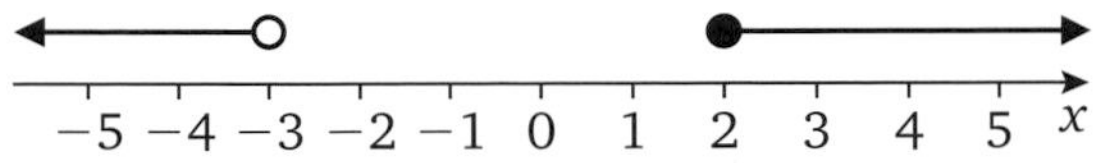

3 −3, −2, −1, 0, 1

4 a $p \leqslant 3$ **b** $q < 4$ **c** $t < 1\frac{1}{2}$

5 a $x = 2, y = -2$ **b** $x = 3, y = -0.4$ **c** $x = 5, y = -1$ **d** $x = 2, y = -5$

6 $a = 20, b = 15, P = 105°, Q = 75°, R = 105°, S = 75°$

7 $5x + 10y = 470$

$x + y = 65$ $x = 36, y = 29$

8 $x = 7, y = -5$ or $x = -5, y = 31$

9 a

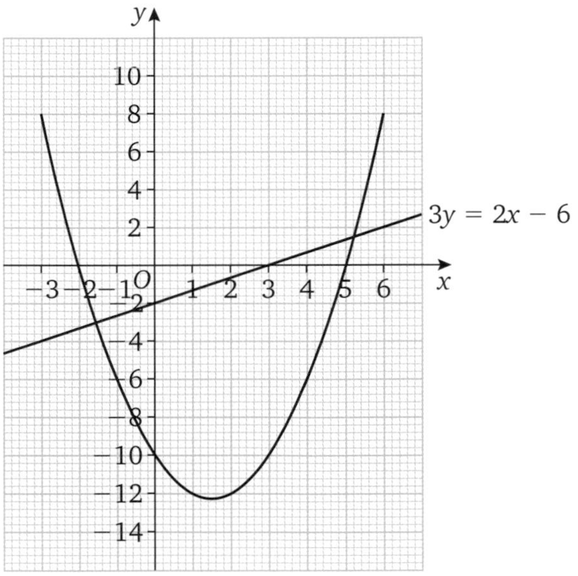

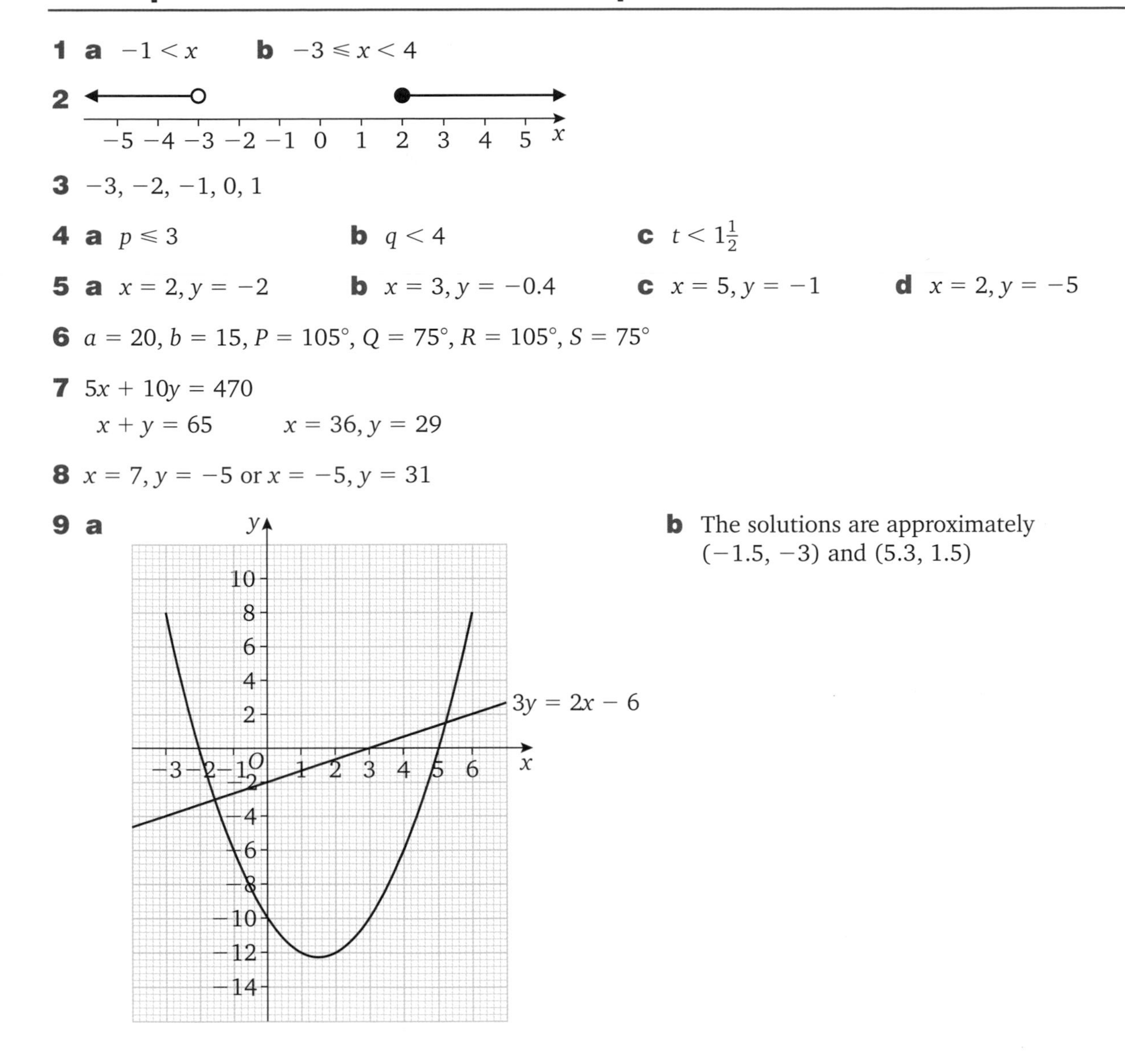

b The solutions are approximately (−1.5, −3) and (5.3, 1.5)

8 AQA Examination-style questions

1 a/b

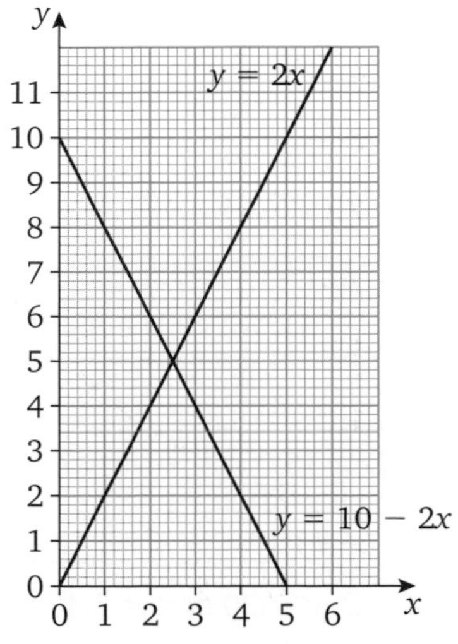

c (2.5, 5)

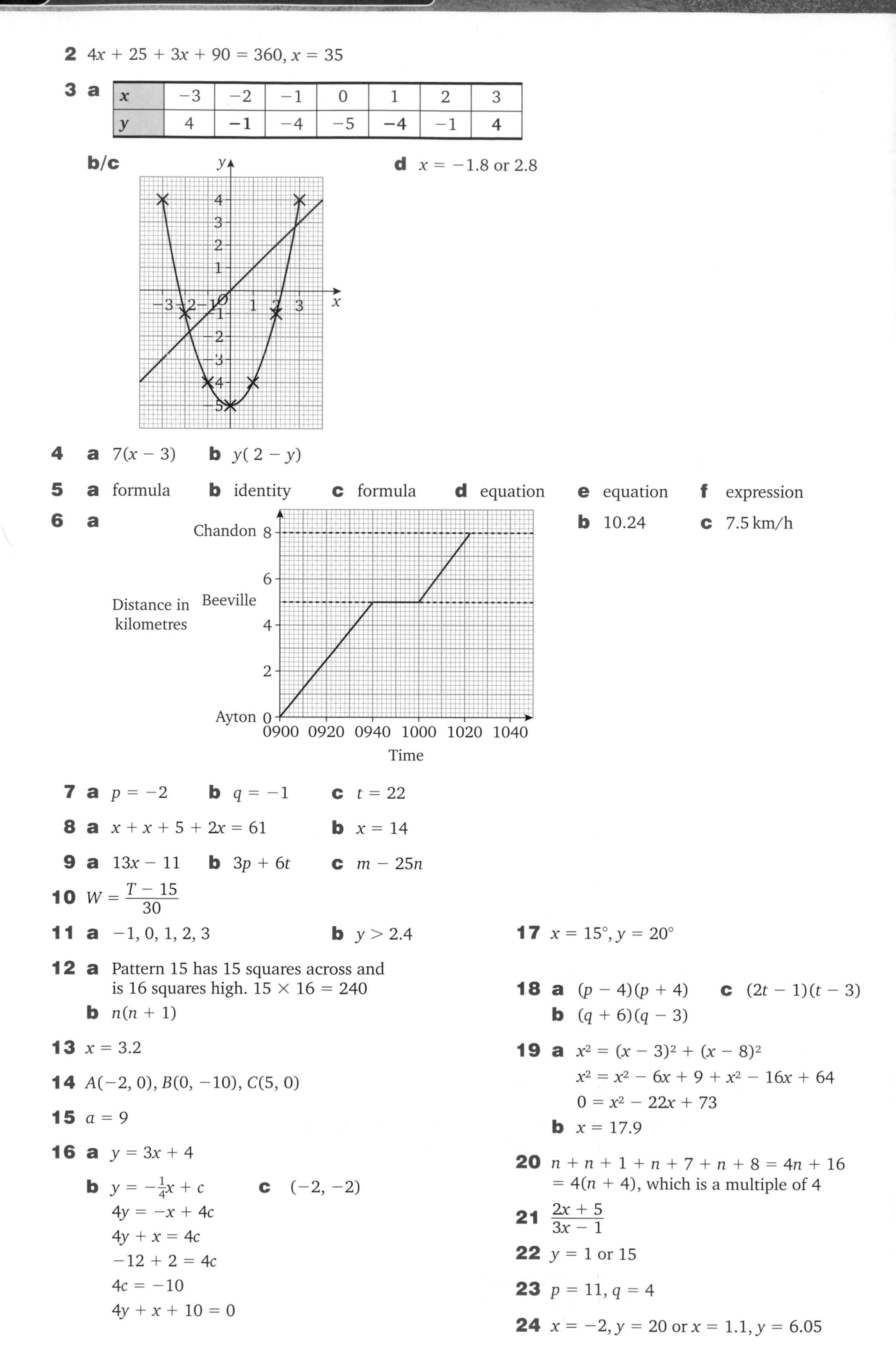

2 $4x + 25 + 3x + 90 = 360, x = 35$

3 a

x	−3	−2	−1	0	1	2	3
y	4	**−1**	−4	−5	**−4**	−1	**4**

b/c

d $x = -1.8$ or 2.8

4 a $7(x - 3)$ **b** $y(2 - y)$

5 a formula **b** identity **c** formula **d** equation **e** equation **f** expression

6 a

b 10.24 **c** 7.5 km/h

7 a $p = -2$ **b** $q = -1$ **c** $t = 22$

8 a $x + x + 5 + 2x = 61$ **b** $x = 14$

9 a $13x - 11$ **b** $3p + 6t$ **c** $m - 25n$

10 $W = \dfrac{T - 15}{30}$

11 a −1, 0, 1, 2, 3 **b** $y > 2.4$

12 a Pattern 15 has 15 squares across and is 16 squares high. $15 \times 16 = 240$

b $n(n + 1)$

13 $x = 3.2$

14 $A(-2, 0), B(0, -10), C(5, 0)$

15 $a = 9$

16 a $y = 3x + 4$

b $y = -\frac{1}{4}x + c$

$4y = -x + 4c$

$4y + x = 4c$

$-12 + 2 = 4c$

$4c = -10$

$4y + x + 10 = 0$

c $(-2, -2)$

17 $x = 15°, y = 20°$

18 a $(p - 4)(p + 4)$ **b** $(q + 6)(q - 3)$ **c** $(2t - 1)(t - 3)$

19 a $x^2 = (x - 3)^2 + (x - 8)^2$

$x^2 = x^2 - 6x + 9 + x^2 - 16x + 64$

$0 = x^2 - 22x + 73$

b $x = 17.9$

20 $n + n + 1 + n + 7 + n + 8 = 4n + 16$
$= 4(n + 4)$, which is a multiple of 4

21 $\dfrac{2x + 5}{3x - 1}$

22 $y = 1$ or 15

23 $p = 11, q = 4$

24 $x = -2, y = 20$ or $x = 1.1, y = 6.05$

Geometry and measure

4 Answers

1 Area and volume

1 Perimeter = 30 cm

2 22.01 cm^2, 30.38 cm^2, 39.71 cm^2, 30.19 cm^2 (to 2 d.p.)

3 3.6 cm (to 1 d.p.)

4 Volume = 2800 cm^3 (to 4 s.f.) and surface area = 1424 cm^2

5 Volume = 1508 cm^3 (to 4 s.f.) and surface area = 1284 cm^2 (to 4 s.f.)

6 Volume of sphere $= \frac{4}{3}\pi r^3 = 36\pi$ cm^3 Volume of cylinder $\pi r^2 h = 54\pi$ cm^3 Fraction occupied $= \frac{36\pi}{54\pi} = \frac{2}{3}$

7 Volume = 59.0 cm^3 (to 1 d.p.)

8 $\theta = 110.5°$ (to 1 d.p.)

9 378π or 1187.5 cm^3 (to 1 d.p.)

2 Angles and polygons

1 **a** 075° **b** 202° **c** 322°

2 **a** 47° (alternate angles on parallel lines)
b 69° (angles on a straight line)
c 64° (alternate angles on parallel lines)

3 45°

4 15 sides

5 A regular hexagon has interior angles of 120°, which fit three times into 360°.
A regular pentagon has interior angles of 108°, which will not fit exactly into 360°.

6 **a** **i** $BCD = 108°$ (angle sum of pentagon $= (5 - 2) \times 180 = 540$; $540 \div 5 = 108$)
ii $EDA = 36°$. ED is isosceles triangle with $AED = 108°$
iii $ADC = 72°$ ($EDC - EDA = 108 - 36$)
b $ADC + BCD$ are allied angles ($72 + 108 = 180°$)

7 $x + x + 2x + x + 10 + 100 = 540$
$5x + 110 = 540$
$5x = 430$
$x = 86$

8 Call the angles x, x, x, x and $2x$. Then $x + x + x + x + 2x = 540$
$6x = 540$
$x = 90$ So the angle of $2x = 180°$ or a straight line.
So the shape is a rectangle.

3 Circle theorems

1 $ACB = 64°$

2 $DAC = DBC = 32°$ (angles in the same segment)
$ADC = 150 - 32 - 58 = 90°$ (angles in a triangle)
AC is a diameter (angle in a semicircle = 90°)

3 a $OAD = 28°$ (alternate segment)

b $OBD = 62°$ (angles in a triangle $= 180°$; $ADB = 90°$, angle in a semicircle)

c $BCD = 34°$ (angles on a straight line $= 180°$)
(DOB $= 56°$, angle at centre $= 2 \times$ angle at circumference, angles in a triangle)

4 a $DAC = 28°$ (angles in the same segment)

b $ADC = 90°$ (angle in a semicircle)

c $ACD = 62°$ (angles in a triangle)

5 $ABC = 120°$

6 $BAD + ADC = 180°$ (allied angles)
$BAD + BCD = 180°$ (opposite angles of cyclic quadrilateral)
So $ADC = BCD$
So the trapezium $ABCD$ is isosceles.

4 Transformations and vectors

1, 2

3 Reflection in $y = 2 - x$

4

5

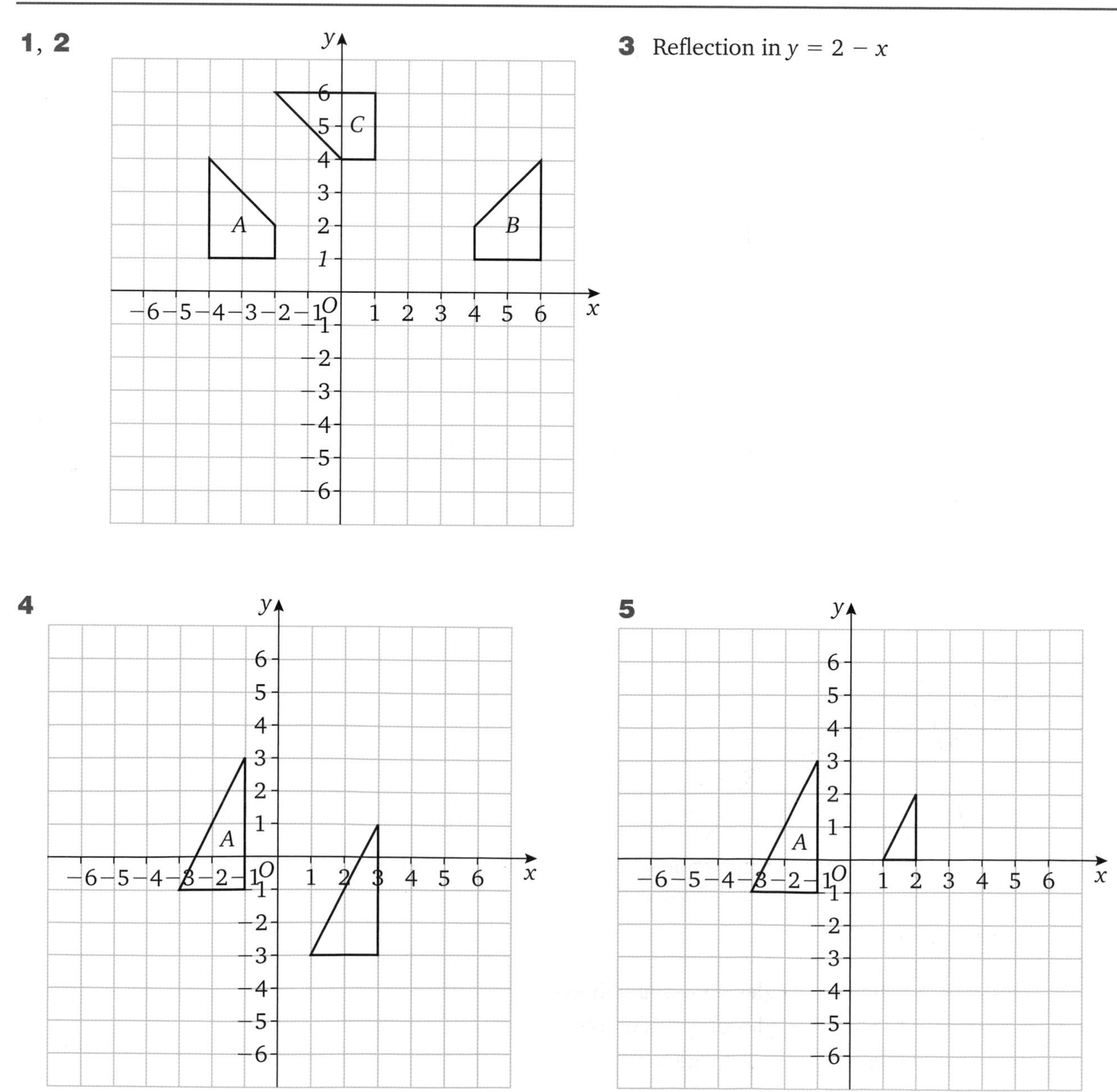

6

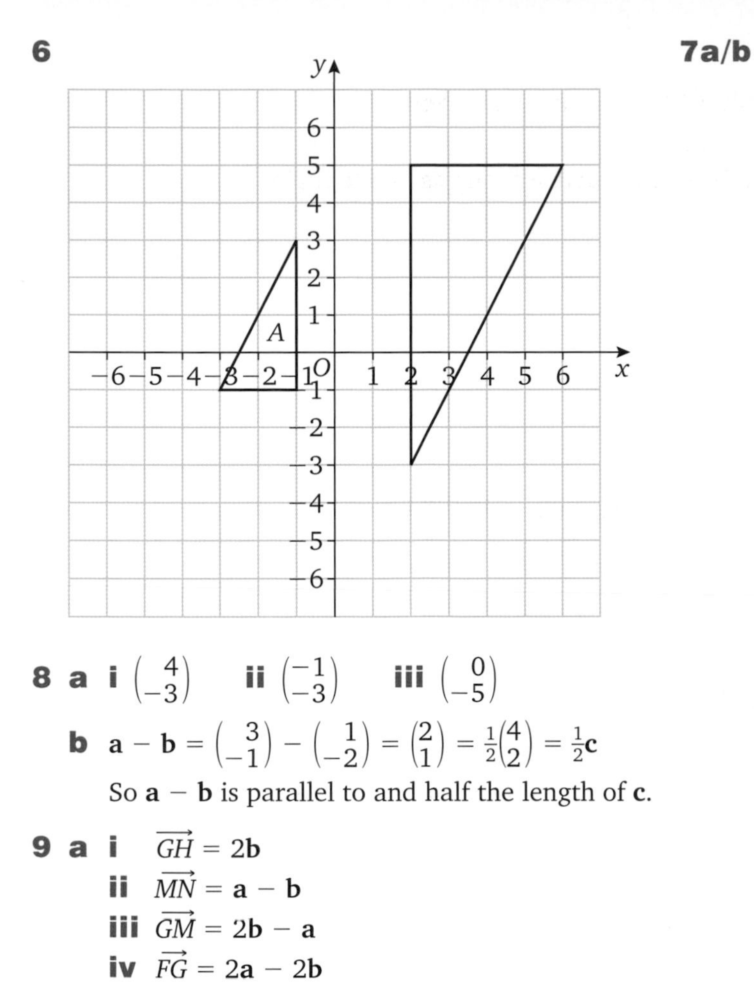

7a/b

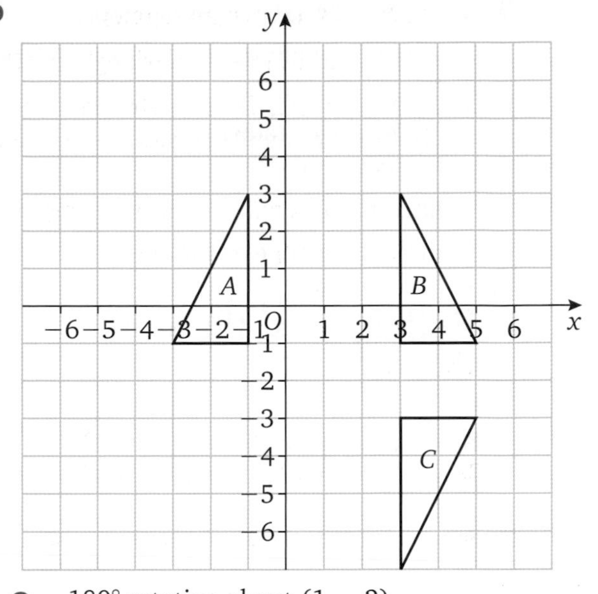

c 180° rotation about (1, −2)

8 a i $\begin{pmatrix} 4 \\ -3 \end{pmatrix}$ **ii** $\begin{pmatrix} -1 \\ -3 \end{pmatrix}$ **iii** $\begin{pmatrix} 0 \\ -5 \end{pmatrix}$

b $\mathbf{a} - \mathbf{b} = \begin{pmatrix} 3 \\ -1 \end{pmatrix} - \begin{pmatrix} 1 \\ -2 \end{pmatrix} = \begin{pmatrix} 2 \\ 1 \end{pmatrix} = \frac{1}{2}\begin{pmatrix} 4 \\ 2 \end{pmatrix} = \frac{1}{2}\mathbf{c}$

So $\mathbf{a} - \mathbf{b}$ is parallel to and half the length of $\mathbf{c}$.

9 a i $\overrightarrow{GH} = 2\mathbf{b}$

ii $\overrightarrow{MN} = \mathbf{a} - \mathbf{b}$

iii $\overrightarrow{GM} = 2\mathbf{b} - \mathbf{a}$

iv $\overrightarrow{FG} = 2\mathbf{a} - 2\mathbf{b}$

b *FG* and *MN* are parallel, with *FG* = 2*MN*

5 Measures, loci and construction

1 Student's own drawing of an equilateral triangle with sides of 6 cm.

2 7.9 m/s (to 1 d.p.)

3

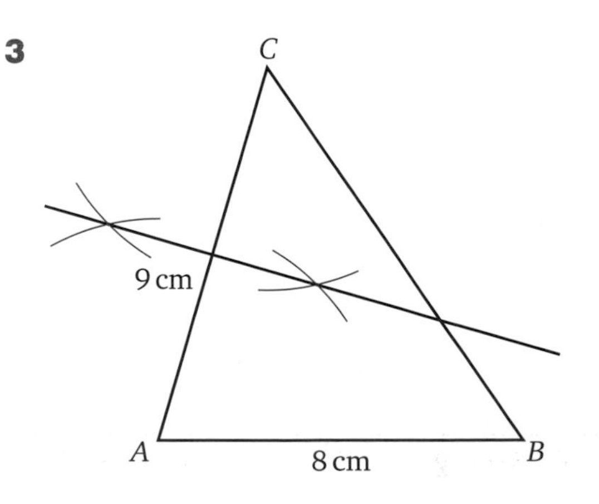

4 7.8 g/cm³ (to 1 d.p.)

5 a *ADE* = *ECB* (alternate angles on parallel lines)

DAE = *EBC* (alternate angles on parallel lines)

AED = *BEC* (vertically opposite angles)

b *DE* = 2.7 cm

6 $BAC = 90°$ (angle in a semicircle)
$OED = 90°$ (angle between tangent and chord)
$BAC = OED = 90°$
$AB = OC = OE$ (radii)
$BO = OC = CD$
So $BC = OD$
Triangles are congruent (RHS).

6 Pythagoras' theorem

1 a 10 cm
b 6.0 cm (to 1 d.p.)
c 7.1 m (to 1 d.p.)

2 4.1 cm (to 1 d.p.)

3 $AB = 5$ units

4 $5^2 + 12^2 = 169 = 13^2$

5 5.66 m (to the nearest cm)

6 8.5 cm (to 1 d.p.)

7 12.4 cm (to 1 d.p.)

7 Trigonometry

1 a $a = 3.8$ cm **b** $b = 42.6°$ **c** $c = 6.3$ cm (all to 1 d.p.)

2 $A = B = 31°$, $C = 118°$

3 $a = 6.2$ cm, $b = 75.7°$ or $104.3°$, $c = 4.3$ cm (all to 1 d.p.)

4 28.2 cm^2 (to 1 d.p.)

5 $XZ = 5.8$ km, $YZ = 5.7$ km (both to 1 d.p.)

6 10.2 cm (to 1 d.p.)

8 AQA Examination-style questions

1 Area of $B = 2 \times$ area of A
Radius of $B = \sqrt{2} \times$ area of $A = 11.3$ cm (to 1 d.p.)

2 Speed = 100 m ÷ 3.9 s = 25.6 m/s = 25.6×60^2 = 92 307.69231 m/h = 92.3 km/h (to 1 d.p.)

3 a $ACB = 90°$ (angle in a semicircle)
$\cos A = \frac{9}{12} = 0.75$
$A = \cos^{-1} 0.75 = 41.4°$
$BOC = 2 \times CAB = 82.8°$ (angle at centre = 2 × angle at circumference)

b By Pythagoras:
$AC^2 + BC^2 = AB^2$
$81 + BC^2 = 144$
$BC = \sqrt{63} = 7.9$ cm (to 1 d.p.)

4 Area $= \frac{1}{2}ab \sin C = \frac{1}{2}64 \sin 60 = 27.7$ cm^2 (to 1 d.p.)

5 Angle $CBA = 211 - 102 = 109°$
$b^2 = a^2 + c^2 - 2ac \cos B$
$b^2 = 12.4^2 + 8.2^2 - 2 \times 12.4 \times 8.2 \cos 109$
$b^2 = 287.2075399$
$b = 16.9$ km (to 1 d.p.)

6 a i $\overrightarrow{AC} = 2\mathbf{a} - 2\mathbf{b}$ **ii** $\overrightarrow{AN} = 2\mathbf{a} - \mathbf{b}$ **iii** $\overrightarrow{MN} = \mathbf{a} - \mathbf{b}$

b $\overrightarrow{XB} = \overrightarrow{XN} + \overrightarrow{NB} = \frac{1}{3}\overrightarrow{AN} + \overrightarrow{NB} = \frac{1}{3}(2\mathbf{a} - \mathbf{b}) + \mathbf{b} = \frac{2}{3}\mathbf{a} + \frac{2}{3}\mathbf{b}$

Answers

1 Problem-solving

1 To take 72p from her pocket Chandi has to take a 50 pence coin, a 20 pence coin and a 2 pence coin.
For this to be the largest possible amount, the other coin must be a 1 pence coin.
So when Chandi takes three coins from her pocket she can take
72p (50 + 20 + 2), 71p (50 + 20 + 1), 53p (50 + 2 + 1), 23p (20 + 2 + 1)

2 $4n - 10 + 2n - 20 + 3n + 25 + n + 15 = 360$
$10n = 350$
$n = 35$
Angle $A = 130°$, angle $B = 50°$, angle $C = 130°$ and angle $D = 50°$
Opposite angles are equal so the quadrilateral is a parallelogram.

3 Rowing machine 12.5 minutes, cross-trainer 25 minutes

4 If x is the number they both think of $4(x - 3) = 2x + 15$
This gives $x = 13.5$

5 25 minutes

6 $6x + 4y = 49$
$8x - 2y = 36$
Solving this gives $x = 5.5$ and $y = 4$
Rectangle A has dimensions 23 cm by 1.5 cm and area 34.5 cm^2.
Rectangle B has dimensions 8.5 cm by 9.5 cm and (largest) area of 80.75 cm^2.

2 Proof

1 $n + (n + 1) + (n + 2) + (n + 3) + (n + 4) \equiv 5n + 10 \equiv 5(n + 2)$
$5(n + 2)$ is $5 \times$ an integer and, therefore, a multiple of 5.

2 **a** $2n$ is $2 \times$ any integer and, therefore, a multiple of 2.
$2n + 1$ is, therefore, 1 more than an even number, and odd.
b $(2n + 1) + (2n + 3) + (2n + 5) \equiv 6n + 9 = 3(2n + 3)$
$3(2n + 3)$ is $3 \times$ an integer and, therefore, a multiple of 3.

3 $(2n + 1)(2n + 3) \equiv 4n^2 + 8n + 3 \equiv 4n^2 + 8n + 4 - 1 \equiv 4(n^2 + 2n + 1) - 1 \equiv 4(n + 1)^2 - 1$
$4(n + 1)^2$ is $4 \times$ an integer and, therefore, a multiple of 4.
So $4(n + 1)^2 - 1$ is 1 less than a multiple of 4.

4 Perimeter $= \frac{x-1}{x+1} + \frac{x-1}{x+1} + \frac{x-3}{x^2-1} + \frac{x-3}{x^2-1}$

$$\frac{x-1}{x+1} + \frac{x-1}{x+1} \equiv \frac{x-3}{x^2-1}\frac{x-3}{x^2-1} = \frac{(x-1)^2 + (x-1)^2 + 2x - 6}{x^2-1}$$
$$\equiv \frac{2(x^2 - 2x + 1) + 2x - 6}{x^2 - 1}$$
$$\equiv \frac{2x^2 - 2x - 4}{x^2 - 1}$$
$$\equiv \frac{2(x+1)(x-2)}{x^2 - 1}$$
$$\equiv \frac{2(x-2)}{x+1}$$

5 If the triangle has a right angle at B, $AB^2 + BC^2 = AC^2$
$AB^2 = x^2$ and $BC^2 = (\sqrt{(2x + 1)})^2 \equiv 2x + 1$
$AB^2 + BC^2 \equiv x^2 + 2x + 1 \equiv (x + 1)^2 = BC^2$

6 $(3x + 2)^2 - (2x + 3)^2 \equiv [(3x + 2) + (2x + 3)]\,[(3x + 2) - (2x + 3)]$
$(3x + 2)^2 - (2x + 3)^2 \equiv (5x + 5)(x - 1) \equiv 5(x + 1)(x - 1) \equiv 5(x^2 - 1)$

7 To prove a shape is a rectangle you need to prove all its angles are 90° and that its lengths are not all equal (otherwise the shape is a square).

The exterior angle of a hexagon = 360 ÷ 6 = 60°

So angle *FAB* = 180 − 60 = 120° (exterior angle + interior angle = 180°)

AB = *AF* so triangle *AFB* is isosceles (the sides of a regular shape are equal).

So angle *ABF* = angle *AFB* = (180 − 120) ÷ 2 = 30° (angle sum of triangle).

Angle *FBC* = angle *BFE* = 120 − 30 = 90°.

The same argument applies to angles *FEC* and *BCE*.

So the angles of *BCEF* are all 90°.

FB ≠ *BC* (triangle *ABF* is isosceles **not** equilateral)

8 Angle *ABE* = angle *DCE* (angle subtended from same arc)

Angle *AEB* = angle *DEC* (vertically opposite angles)

AE = *ED* (given)

Triangles *AEB* and *DEC* are congruent (AAS).

9 Let angle *PCB* = $x°$

So angle *BAC* = $2x°$ (given)

Angle *ABC* = $(180 - 2x) \div 2 = 90 - x$ (base angle of isosceles triangle)

Sum of angles in triangle *CBP* = 180°

So $x + 90 - x$ + angle *CPB* = 180

So angle *CPB* = 90°

So *CP* is perpendicular to *AB*.

10 a Using the difference of two squares

$(x + n)^2 - (x - n)^2 \equiv [(x + n) + (x - n)]\,[(x + n) - (x - n)] \equiv 2x \times 2n \equiv 4nx$

Using the expansion of brackets

$(x + n)^2 - (x - n)^2 \equiv (x^2 + 2nx + n^2) - (x^2 - 2nx + n^2) \equiv 4nx$

b $\dfrac{x + n}{x - n} - \dfrac{x - n}{x + n} \equiv \dfrac{(x + n)^2 - (x - n)^2}{(x - n)(x + n)} \equiv \dfrac{4nx}{x^2 - n^2}$

3 AQA Examination-style questions

1 Jerry's cuboid has dimensions 15 by 1 by 1 and surface area 62 cm².

Tom's cuboid has dimensions 5 by 3 by 1 and surface area 46 cm².

2 a Student's own scatter graph.

b value from line of best fit

3 9th term = $13a + 21b$

3rd term = $a + b$

6th term = $3a + 5b$

9th term − 3rd term = $12a + 20b = 4(3a + 5b) = 4 \times$ 6th term

4 a D = 84 mm, h = 84 mm

b Surface area of can **A** = $10\,730\pi$ and surface area of can **B** = 7056π
$7056 \div 10\,730 \times 100 = 65.7...$ which is approximately $\frac{2}{3}$

5 *YZ* = *YZ* (common)

Angle *NZY* = angle *MYZ* (given)

Angle *NYZ* = angle *MZY* (base angles of an isosceles triangle)

So triangles *YNZ* and *YMZ* are congruent (ASA).

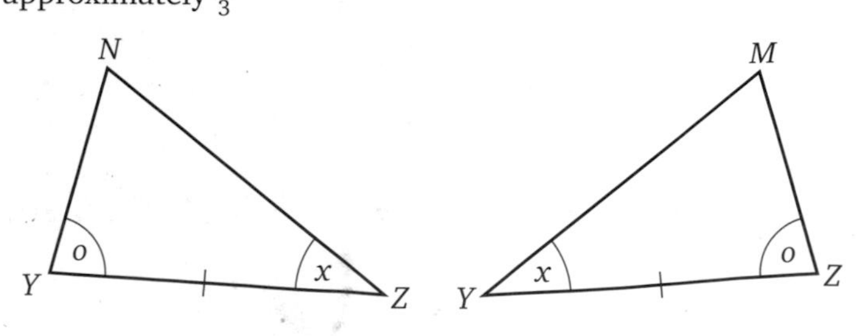